Springer Theses

Recognizing Outstanding Ph.D. Research

For further volumes:
http://www.springer.com/series/8790

Aims and Scope

The series "Springer Theses" brings together a selection of the very best Ph.D. theses from around the world and across the physical sciences. Nominated and endorsed by two recognized specialists, each published volume has been selected for its scientific excellence and the high impact of its contents for the pertinent field of research. For greater accessibility to non-specialists, the published versions include an extended introduction, as well as a foreword by the student's supervisor explaining the special relevance of the work for the field. As a whole, the series will provide a valuable resource both for newcomers to the research fields described, and for other scientists seeking detailed background information on special questions. Finally, it provides an accredited documentation of the valuable contributions made by today's younger generation of scientists.

Theses are accepted into the series by invited nomination only and must fulfill all of the following criteria

- They must be written in good English.
- The topic should fall within the confines of Chemistry, Physics, Earth Sciences, Engineering and related interdisciplinary fields such as Materials, Nanoscience, Chemical Engineering, Complex Systems and Biophysics.
- The work reported in the thesis must represent a significant scientific advance.
- If the thesis includes previously published material, permission to reproduce this must be gained from the respective copyright holder.
- They must have been examined and passed during the 12 months prior to nomination.
- Each thesis should include a foreword by the supervisor outlining the significance of its content.
- The theses should have a clearly defined structure including an introduction accessible to scientists not expert in that particular field.

Dario Villamaina

Transport Properties in Non-Equilibrium and Anomalous Systems

Doctoral Thesis accepted by
the Universitá La Sapienza, Italy

Author
Dr. Dario Villamaina
LPTMS
Université of Paris-Sud
Orsay Cedex
France

Supervisors
Prof. Angelo Vulpiani
Dr. Andrea Puglisi
Dipartimento di Fisica
Universitá La Sapienza and CNR-ISC
Rome
Italy

ISSN 2190-5053 ISSN 2190-5061 (electronic)
ISBN 978-3-319-34980-0 ISBN 978-3-319-01772-3 (eBook)
DOI 10.1007/978-3-319-01772-3
Springer Cham Heidelberg New York Dordrecht London

Softcover reprint of the hardcover 1st edition 2014

Printed on acid-free paper

Springer is part of Springer Science+Business Media (www.springer.com)

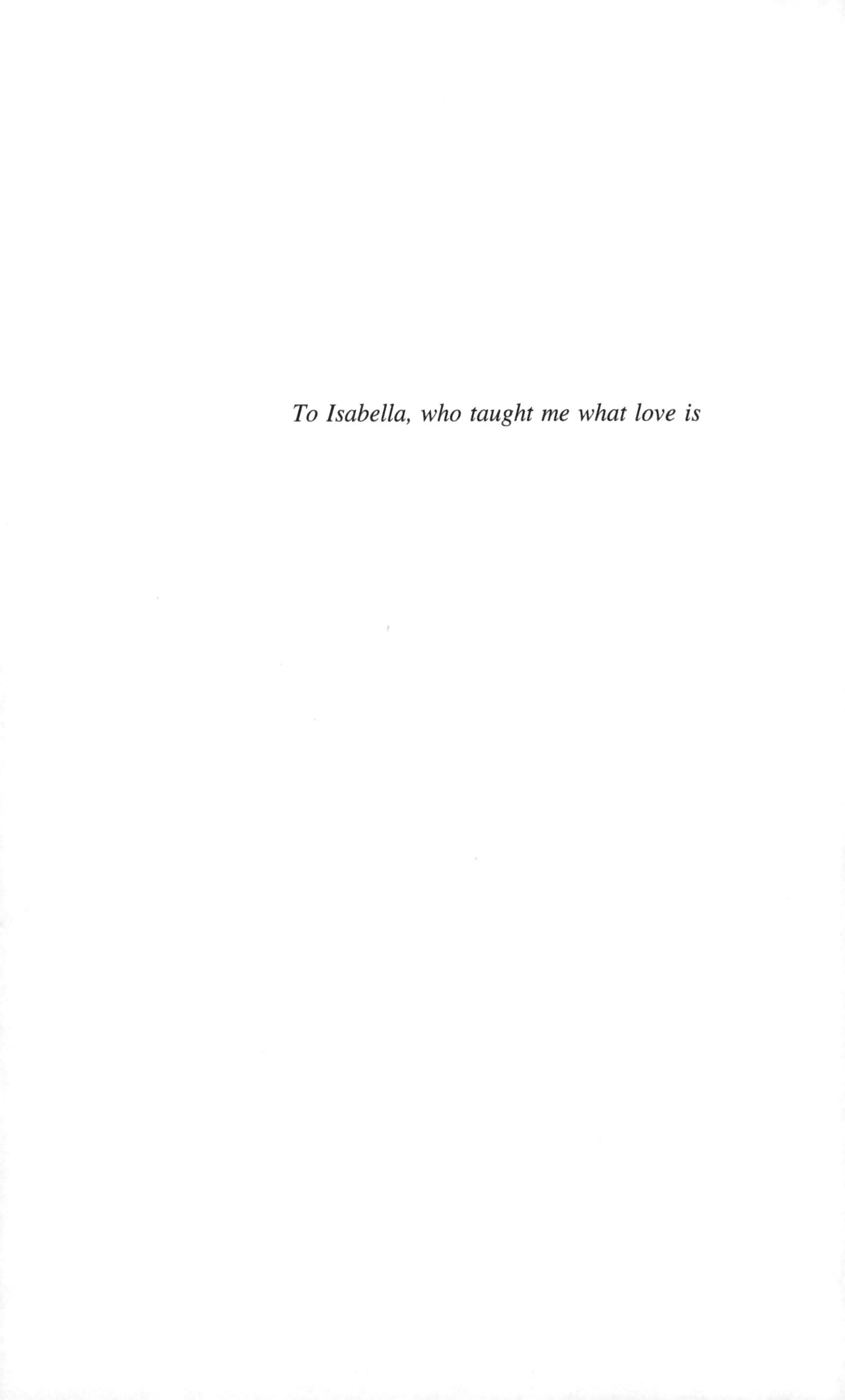

To Isabella, who taught me what love is

Supervisors' Foreword

The study of fluctuations in statistical physics has a long history, starting from Einstein's theory, who was able to connect the fluctuations of macroscopic observables to the entropy. Later, in his seminal paper, Lars Onsager pointed out in a general context the deep relation between the relaxation of an external perturbation and the equilibrium fluctuations. A first example of a fluctuation/dissipation theorem had been derived by Einstein in his seminal work on Brownian motion, where the celebrated relation between mobility and diffusion was shown to be valid. The fluctuation/dissipation theorem and the linear response theory are fundamental aspects of statistical mechanics and have brought many important applications.

In the last decades, the development of modern technology has made possible the study of systems at the nanoscale, where fluctuations play a relevant role, ranging from the physics of colloids and small systems to biological inspired experiments, like single molecule manipulations and molecular motors. Quite surprisingly, when nonconservative forces or non-thermal fluctuations are present, a "simple" approach based on the standard equilibrium statistical mechanics fails to provide correct predictions for the observables of interest. Up to now a general theory of non-equilibrium fluctuations is still an open and challenging problem, being the main point that inspired the research presented in this book.

The work of Dr. D. Villamaina has the ambition to contribute toward the construction of such a general theory, by adding new interesting pieces to the puzzle. His thesis leads the reader from the analysis of simple models of heat conduction to the more realistic cases of granular materials and systems exhibiting anomalous transport. In all these cases the limits of the standard equilibrium approaches are discussed and the proper non-equilibrium fluctuation theory is derived. From the analysis of the different discussed systems, a general lesson does emerge: the correlations among different degrees of freedom can work as a channel of energy exchange, being responsible for the breaking of the equilibrium fluctuation theory. By following this line of reasoning, the author sheds light on the main source of irreversibility in the granular gases, being the cross correlations among the velocities of different particles. This theoretical prediction has also been experimentally verified in a series of recent publications.

In summary, the work of Dr. D. Villamaina shows in a convincing way how the study of non-equilibrium fluctuations and their relation with the response to an external perturbation are crucial for the construction of a proper and effective general theory. The research presented here can be extended to different classes of non-equilibrium and anomalous systems. In conclusion we are sure that this thesis will stimulate the debate, attracting new students and increasing the interest in this research field where many questions are still to be answered.

Rome, June 2013

Prof. Angelo Vulpiani
Dr. Andrea Puglisi

Acknowledgments

I warmly thank Andrea and Angelo who have directed and strongly supported my Ph.D. thesis. They are my mentors. Their guidance and friendship taught me much more than could ever be expressed in words or calculations. I give credit to Alessandro and Giacomo, not only for our research collaboration but also for the nice time spent out of university. I will never forget their encouragement during the drafting of this thesis. Fabio Cecconi taught me the perseverance necessary to work out physics problems. I mention also Andrea Crisanti with pleasure for the countless discussions on physics and his patience in explaining to me all the tricky details of our common work, now part of this thesis. Last but not least, I could always count on Pierpaolo Vivo, who has been always ready for any kind of help and advice. A big "thank you!" to all of them.

I acknowledge my family, that always supported my choices.

Contents

Chapter 1
Introduction

Abstract In this chapter the main open issues that are behind this work are presented. In the first part it is explained why the standard tools of equilibrium statistical mechanics fail to give correct predictions when some non-equilibrium current is present. The second part is devoted to briefly summarize the models of granular material and anomalous transport handled in this work.

In the last century, equilibrium statistical mechanics succeded in describing a huge class of phenomena. The main reasons for this success and the generality of this theory reside mainly in two aspects. First, the application of the ensemble theory to a model requires, as input, only its Hamiltonian and not all the dynamical details. In a sense, when ensemble theory can be applied, all the dynamical properties can be predicted by averaging over the well known Boltzmann-Gibbs distribution, which is valid for a large variety of systems at equilibrium. On the other hand, the thermodynamical quantities can be straightforwardly deduced. Such a connection with the macroscopic world is at the origin of the predictive power of statistical mechanics: from the microscopic interaction rules the emerging thermodynamical level is simply deduced and well-funded. If, from a statical point of view, the main common feature is given by the Boltzmann-Gibbs distribution, from a dinamical point of view the detailed balance condition, namely the symmetry with respect to the time inversion, plays a crucial role. From such a symmetry one can deduce one of the most important results valid at equilibrium: the fluctuation-dissipation theorem, which links the response function to an external perturbation with a simple correlation computed in the equilibrium state. The strength of this relation resides also in its semplicity. As example, it is sufficient to cite the Einstein relation between mobility and diffusion: all the information content of the response to a force acting on a tracer particle is in its autocorrelation, and the other degrees of freedom like the velocity and the position of the surroinding particles are not involved.

Remarkably, the scenario described above falls enterily down as soon as some current is flowing across the system, driving it out of equilibrium. The presence of currents is quite common in nature and produces a richness of phenomena which are far from being included in a general framework. In this work we mainly consider

D. Villamaina, *Transport Properties in Non-Equilibrium and Anomalous Systems*, Springer Theses, DOI: 10.1007/978-3-319-01772-3_1,

systems which are supposed to be ergodic and reach a statistical stationary state in physically observable times. Generally, in these cases, the energy lost by dissipative forces is balanced by some mechanism of external energy injection. Because of dissipation, these systems present an arrow of time, even in the steady state: the probability of a trajectory in phase space is in general different from the one obtained by inverting the direction of the time. This is equivalent to say that detailed balance condition is broken. As a consequence, the steady state is characterized by an invariant measure that is necessarily not the Boltzmann-Gibbs one and equilibrium statistical mechanics cannot be applied. Moreover the breaking of the detailed balance condition implies a violation of the fluctuation dissipation theorem as it is known at equilibrium.

Despite of these difficulties, in the last decades some quite general results on out-of-equilibrium systems have been established. First, also when detailed balance is broken, it is still possible to find a relation between the response and a suitable correlation computed on the unperturbed system, even if it is not the correlation predicted by the equilibrium fluctuation dissipation theorem. On the other side, in the last decades an observable, known in literature with the name of entropy production, has been proposed as a sort of measure of non equilibrium. In connection with this observable, a class of fluctuation relations has been derived, expressing a symmetry of the Cramer function ruling the large deviation property of the entropy production rate, valid for a huge variety of non equilibrium systems and, in a sense, giving a more complete view of the second law of thermodynamics. Altrough the results like the fluctuation theorem and the generalized response relations are very general, they remain at a microscopic dynamical level and the connections with macroscopic observables are far from being explained. In other words, these tools have a large range of applicability but they lead to conclusions which are still model-dependent.

A possible strategy to tackle the "non equilibrium problem" is to study a simplified and analytically tractable model and then to try to deduce some general lessons. Within this purpose, the starting point of this work is the well known generalized Langevin equation. In the above model both the friction and the noise have the same origin given by the interaction with the surrounding medium and, as shown by Kubo, there is a relation between them which is equivalent to assume an equilibrium dynamics. It appears natural to relax this condition in order to mimick a coupling with more than one energy source, and non equilibrium effects are expected to emerge. In this model, both the response properties and the entropy production can be studied with an analytical approach, showing that a non-trivial coupling between different degrees of freedom plays a crucial role, as soon as non equilibrium conditions are explored.

In order to evaluate the strength of this interpretation, it is necessary to make a comparison with a more realistic system; granular gases in the steady states are ideal for this purpose. A granular gas is mainly costituted by macroscopic grains interacting each other with inelastic collisions. We will consider the driven scase, where the energy dissipated because of inelastic collisions is balanced by an external source and a non equilibrium steady state is reached. This model has been widely studied in literature in different contests. In our work, the starting point is the dynamics of a massive intruder in a dilute regime. We will show that, in this regime, no memory effects are present, and this fact has two important consequences. First, if one observes only one particle, an inelastic collision can be mapped to an elastic one by changing the mass of the second particle. Second, the inelastic collisions due to the other particles act on the same time scale of the external bath: then these two effects are physically indistinguishable. Because of this peculiarities, the system can be considered at equilibrium. In this case we will see that the projection operation used to deduce the dynamics of the intruder cancels all the non equilibrium currents. From this limit case one can argue that, in order to obtain non equilibrium effects, it is necessary to break the molecular chaos approximation, by studying denser systems where one cannot treat the particles as independent, because of recollision effects.

Unfortunately, a general theory derived by the linear Boltzmann equation is still lacking, then, as first extension, a generalized Langevin equation is considered. All the techinques developed for this kind of equations will be transferred in this granular context. Such a phenomenological model allows one to identify the presence of an emerging velocity field, coupled to the intruder and responsible of the non equilibrium effects observed.

In order to test the generality of the above results, the last part of this work is devoted to the study of similar non equilibrium conditions in presence of transport anomalies, namely in presence of subdiffusion. Such an anomaly can emerge for instance when there are strong geometrical constraints acting on the interacting particles. A study of some subdiffusive models will show the non-trivial interplay between non equilibrium conditions and anomalous dynamics.

Anomalous dynamics can emerge also in the presence of disorder, for instance in the context of glassy dynamics. This case is completely different from the others studied in this work: this kind of systems, under certain conditions, is not able to equilibrate because of an extreme slowing down of the relaxation time and exhibits aging. Moreover the cage effect induces strong constraints and an intruder exhibits subdiffusive beaviour. Despite of this differences, we will show how the ratchet effect, tipical of stationary and diffusive non equilibrium systems, can be explored also in a glassy phase.

The work is divided in the following chapters:

- In Chap. 2 a brief collection of the results present in literature and used in this work is described. We start with a derivation of the Langevin equation in a way that makes clear the assumptions on the basis of equilibrium dynamics. Then, generalized response relations are presented and the role of entropy production is discussed. Since a large part of this work regards the study of granular gases, the second part of the chapter is entirely devoted to them, paying attention to the still open problems in dense regimes. This is not a chapter of a review article, and for this reason it could appear incomplete. However, it must be seen as an occasion to present the common ground where there are the basis of our research, and it proposes some questions which are developed and, at least partially, solved in the rest of the work.
- In Chap. 3 the Langevin equation with memory is analyzed in both equilibrium and non equilibrium setup. Non-Markovian equation can be mapped in a Markovian

one by increasing enough the number of degrees of freedom. This procedure is not just a simple mathematical trick, on the contrary the relative coupling between different variables is relevant for the correct prediction of the response. Moreover such a coupling is proportional to the entropy production rate. By concluding, we show how the mapping from Markovian to non-Markovian dynamics is equivalent to a projection operation and it carries a loss of information that can be detected by entropy production.

- Chapter 4 is entirely devoted to a granular gas model. A Langevin equation for a massive tracer is obtained from the linear Boltzmann equation via a Kramers-Moyal expansion in a diluite limit. Such an expansion is not sufficient to observe non equilibrium effects and an equilibrium-like effective regime is obtained, without the presence of memory. In a denser regime, when the molecular chaos fails, the equation for a tracer is well represented by a Langevin equation with memory, and a local velocity field plays the role of an auxiliary variable coupled to the tracer. Finally, a strong assessment of the validity of the "local field" interpretation is given by the numerical verification of the fluctuation relation.
- In Chap. 5 the additional ingredient of anomalous diffusion, combined with non equilibrium conditions, is studied. The analysis starts with a random walk on a comb lattice, which can be analitically solvable. A detailed analysis of the "single file model" is then shown and the response analysis is similar to that one in higher dimensions. The chapter ends with the study of a ratchet effect in a fragile glass former. Because of the presence of disorder, under certain conditions, an intruder in a glass former exhibits subdiffusion. Despite of the great differences from the "family" of the non equilibrium steady states, also in this system it is possible to observe a ratchet effect, although characterized by a subvelocity due to the disorder.

Finally, some conclusions are drawn in Chap. 6.

Chapter 2
Non-Equilibrium Steady States

Abstract In this chapter a brief collection of the results present in literature and used in this work is described. We starts with a derivation of the Langevin equation in a way that makes clear the assumptions on the basis of equilibrium dynamics. Then, generalized response relations are presented and the role of entropy production is discussed. Since a large part of this work regards the study of granular gases, the second part of the chapter is entirely devoted to them, paying attention to the still open problems in dense regimes. This is not a chapter of a review article, and for this reason it could appear incomplete. However, it must be seen as an occasion to present the common ground where there are the basis of our research, and it proposes some questions which are developed and, at least partially, solved in the rest of the work.

The fluctuations of a system at equilibrium are characterized by strong symmetries that connect dissipation and noise. In this way, it is possible to have a prediction of the response to an external perturbation by only measuring a coniugated correlation function in the unperturbed system. As soon as non conservative forces are present, this symmetries are broken. In this case, the complete characterization of fluctuation and response is still an open problem of statistical mechanics.
In this chapter we start from a derivation of the Langevin equation from first principles, as example of equilibrium dynamics. From this example it appears evident that the fluctuating part and the dissipative part are both linked to the projection of fast degrees of freedom and are then connected. Using this case as a starting point, we present a derivation of the generalized response relations and the fluctuation relations for generic stochastic processes. Some specific phenomena typical of non-equilibrium states are then deduced from this general theory, like the connection between entropy production and the arrow of time, the ratchet effect of an asymmetric object in a non-equilibrium environment and finally the concept of effective temperature for aging systems. In the last part of the chapter, granular gases are introduced. In this class of models the presence of dissipation due to inelastic collisions, when balanced by an external energy injection mechanism, produces a striking example of non-equilibrium steady state. In granular gases, the description of non-thermal

D. Villamaina, *Transport Properties in Non-Equilibrium and Anomalous Systems*, Springer Theses, DOI: 10.1007/978-3-319-01772-3_2,

fluctuations, the response mechanism and the entropy production are still unclear from a theoretical point of view. These open questions are here briefly presented in a general context, and the search for a proper answer will be the central issue in the rest of the work.

2.1 Historical Notes: The Central Role of the Fluctuations

The study of fluctuations has a great importance in statistical mechanics. Historically, it is common and appropriate to start from the work of the botanist Robert Brown [16]. In 1827, by using a microscope, he observed grains of pollen of the plant *Clarkia pulchella* suspended in water moving in a very irregular way. Contrary to the common thinking, Robert Brown was not the first one to discover the Brownian motion (in a paper [17] he mentions several precursors) but his main contribution was to unveil the pure mechanical origin of this phenomena. As written in a review of that period [18]:

This motion certainly bears some resemblance to that observed in infusory animals, but the latter show more of a voluntary action. The idea of vitality is quite out of the question. On the contrary, the motions may be viewed as of a mechanical nature, caused by the unequal temperature of the strongly illuminated water, its evaporation, currents of air, and heated currents...

Thirty years after the work of Brown, the French physicist Louis Georges Gouy, supporting kinetic theory, pointed out several peculiarities of this motion, as reported in Perrin's book [73]. Among others, the most relevant are:

- the motion is very irregular, it appears that the trajectory has no tangent, and close particles move in independent way
- by increasing the temperature of the solvent, the motion is "more active"
- the motion never ceases or change qualitatively.

These features could be explained via kinetic arguments, and a direct test of it resides in the equipartition law. However, before the celebrated Einstein's work, several experimentalists failed to estimate the velocity of the tracer particle because of its irregularity and confirmation of kinetic theory was not possible (see [67] and references therein).

The breakthrough in understanding this phenomena arrived independently from Smoluchowski [81] and Einstein [29]. The conceptual relevant point of the work of Einstein is the assumption of the statistical equilibrium of the particle with the surrounding medium, together with the Stokes law experienced by a particle immersed in a fluid.

Based on this intuition, Langevin [58] proposed a stochastic differential equation for the velocity of a Brownian particle:

$$\frac{dV}{dt} = f_s + \xi(t) = -\gamma V + \xi(t), \tag{2.1}$$

where f_s is the Stokes law with $\gamma = 6\pi\eta a$, a is the radius of a molecule, η is the viscosity and $\xi(t)$ is a fluctuating force, whose variance is fixed by the equipartition law. The importance of fluctuations is now clear: a computation with only the Stokes term would produce an exponential relaxation and no movement would be predicted. From (2.1) an expression for the diffusion coefficient is obtained

$$D = \lim_{t\to\infty} \frac{\langle [x(t) - x(0)]^2 \rangle}{2t} = \frac{RT}{6N_A\pi\eta a}, \tag{2.2}$$

where N_A is the Avogadro number and R is the gas constant. The evocative aspect of Eq. (2.2) is given by the possibility of counting molecules, by observing the macroscopic fluctuations of the position $x(t)$ of a tracer particle, whose measure is clearly easier than velocity estimations, as tried in the past.

This relation was experimentally confirmed by Svendberg and Perrin, dispelling any doubt on the atomic theory.

On the other side, the dynamical equations introduced by Langevin have a wide range of applicability and have been generalized and deduced in several contexts.

2.1.1 The Origin of the Langevin Equation: Noise and Friction

Let us consider a system coupled to a thermal bath, for example a massive intruder in a fluid, and let us suppose we are interested in obtaining its dynamical equations. The basic idea is to start from a full description of the variables present in the system, and then to obtain an effective dynamical equation by reducing the number of degrees of freedom. Generally speaking, the Hamiltonian of the system can be split into three parts:

$$H_{tot} = H_{system} + H_{bath} + H_{int}, \tag{2.3}$$

where H_{int} is the interaction term. A standard projection recipe consists in integrating over the bath variables and then in obtaining some dynamical equations for the "slow" variables of the system of interest. In this section, we will describe, as particular case, a harmonic model introduced for the first time by Zwanzig [88], which has the advantage of being analytically tractable. In this case one has

$$H_{system}(X, P) = \frac{P^2}{2M} + U(X) \tag{2.4}$$

$$H_{bath} + H_{int} = \sum_j \left[\frac{p_j^2}{2} + \frac{1}{2}\omega_j^2 \left(x_j - \frac{\gamma_j}{\omega_j^2} X \right)^2 \right]. \tag{2.5}$$

where the capital letters refer to the tracer particle and the bath is described by the collection of variables $\{x, p\}$. Note that the strength of the interaction term is ruled

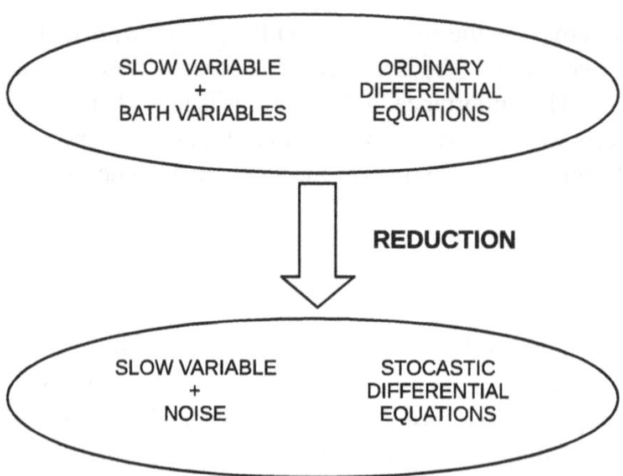

Fig. 2.1 Scheme of the reduction process

by γ_i. The equations of motion read

$$M\frac{dX}{dt} = P \quad \frac{dP}{dt} = -U'(X) + \sum_j \gamma_j (x_j - \frac{\gamma_j}{\omega_j^2} X) \tag{2.6}$$

$$\frac{dx_j}{dt} = p_j \quad \frac{dp_j}{dt} = -\omega_j^2 x_j + \gamma_j X. \tag{2.7}$$

Now, thanks to the harmonic choice of the bath variables, it is evident that the Eq. (2.7) can be integrated and substituted in (2.6), yielding an equation for the variables (P, X) depending only on the initial conditions $\{q(0), p(0)\}$. The equation for the variable P can be recast into:

$$\frac{dP}{dt} = -U'(x) - \int_0^{+\infty} ds K(s) \frac{P(t-s)}{M} + F_p(t), \tag{2.8}$$

where

$$K(t) = \sum_j \frac{\gamma_j^2}{\omega_j^2} \cos \omega_j t \tag{2.9}$$

$$F_p(t) = \sum_j \gamma_j p_j(0) \frac{\sin \omega_j t}{\omega_j} + \sum_j \gamma_j \left(q_j(0) - \frac{\gamma_j}{\omega_j^2} x(0) \right) \cos \omega_j t. \tag{2.10}$$

Up to now, no approximation has been done: these equations are indeed a simple rewriting and the level of the description is still Hamiltonian, with a deterministic evolution depending on the initial conditions. Clearly, for a large numbers of oscillators, for instance of the order of the Avogadro number, referring to the initial condition in order to maintain the deterministic nature of the analysis is ingenuous and useless, and a probabilistic approach is necessary.

The statistical ingredient in the description is implemented by considering an equilibrium canonical distribution at a well defined temperature[1] $T = \frac{1}{\beta}$ for the initial conditions of the bath oscillators:

$$\rho(x, p) \propto e^{-\beta(\mathcal{H}_{bath}+\mathcal{H}_{int})}. \tag{2.11}$$

The statistical averages of the initial conditions are

$$\left\langle \left(x_j(0) - \frac{\gamma_j}{\omega_j^2} X(0) \right)^2 \right\rangle = \frac{T}{\omega_j^2} \qquad \langle p_j(0)^2 \rangle = T, \tag{2.12}$$

and clearly the first moments and the cross correlation vanishes. With this operation the scenario changes completely, and from a deterministic approach one passes to a stochastic one. As a consequence, the variable $F_p(t)$ depends on the initial condition of the bath, and plays the role of a noise [25]. A central relation in this model is given by:

$$\langle F_p(t) F_p(t') \rangle = T K(t - t'), \tag{2.13}$$

which is called fluctuation dissipation relation of the second kind [50, 52]. Let us conclude this example with some remarks. Thanks to the peculiar form of H_{int}, one obtains that the correlation of the noise does not depend on x. It is possible to show, indeed, that if one introduces a non linear coupling term between the variable X and the bath variables, a multiplicative noise term appears [38]. In some non linear cases, like in some pure kinetic models, some approximations must be taken into account, like the large mass of the intruder, inducing some time scales separation. We will return largely on this point in Chap. 4 with the work, among the others, of Mori and Zwanzig [66, 89], the theory of the Brownian motion and of the generalized Langevin equation has been extended to slow observables, via a projection technique, under very general hypothesis. A central aspect is that, also in these more general approaches, the proportionality between the correlation of noise and the memory term is always verified. One must notice that, as evident from this simple example, both the noise and the memory have the same origin and, as a consequence, a relation connecting them is expected. We will point out in Chap. 3 that (2.13) is substantially equivalent to have taken equilibrium conditions.

In this work we will focus on the classical aspect of non-equilibrium statistical mechanics but it is worth to mention that extensions to the quantum or relativistic case have been developed [28, 39].

---

[1] In this thesis we always measure the temperature in scales of energy, namely we set the Boltzmann constant k_B equal to one.

2.2 General Aspects of Non-Equilibrium Steady States

2.2.1 The Linear Response Relations

The study of the response properties plays a central role in this work. Historically, response theory has been developed first in equilibrium, namely for system described by Hamiltonian dynamics or where the ensemble theory is correct. For this reason, in quite all the textbooks, response theory is presented as a synonymous of the so-called fluctuation dissipation theorem.[2] A consequence of this important theorem was anticipated by Lars Onsager. With the regression hypothesis, he argued that a system cannot "know" if a small fluctuation from equilibrium is caused by an internal fluctuation or by an external field: as a consequence the regression of microscopic thermal fluctuations at equilibrium follows the macroscopic law of relaxation of small non-equilibrium disturbances [70]. Actually there is no apparent reason to apply this "causality principle" only to equilibrium systems: it is possible, indeed, to define the response of a system at a more general level [64], and as we will see, it is always possible to connect it to a suitable correlation. At equilibrium, it assumes well known and tractable forms.

In order to fix ideas, let us suppose that some noise is present. Therefore we consider cases in which it is possible to associate a probability to the trajectories. Let us then consider the space $\{\omega\}$ of trajectories of length t and its probabilities $P_0(\omega)$. Let us consider the effect of an external perturbation: it changes the dynamics and the relative probability of the trajectories in $P_h(\omega)$ (for simplicity in what follows we consider that the space of perturbed trajectories $\{\omega\}$ remain the same). The average value of any observable in presence of the perturbation is easy computed as[3] $\langle O(t)\rangle_h = \sum_\omega O(\omega)P_h(\omega)$. Within this definition, the following identity trivially holds:

$$\langle O(t)\rangle_h = \left\langle O(t)\frac{P_h(\omega)}{P_0(\omega)}\right\rangle_0, \tag{2.14}$$

where, with $\langle \dots \rangle_0$ we denote the average over the unperturbed trajectories. By taking the functional derivative with respect to $h(t')$, the response function is easily obtained:

$$\overline{\frac{\delta O(t)}{\delta h(t')}}\bigg|_{h=0} = \left\langle O(t)\frac{1}{P_0(\omega)}\frac{\delta P_h(\omega)}{\delta h(t')}\bigg|_{h=0}\right\rangle_0, \tag{2.15}$$

where we have introduced $\overline{(\dots)} \equiv \langle \dots \rangle_h$, in order to lighten notation. Equation (2.15) is, as anticipated above, a generalization of the Onsager's sentence, for a general system: the response of an observable to an external perturbation is equal to a suitable correlation computed in the unperturbed system. As it appears

[2] We will omit the expression "of the first kind" When the kind is not specified we always refer to this relation.

[3] We consider a numerable set of trajectories for simplicity of notation.

clear, Eq. (2.14) and its linearized version (2.15) are somehow too general: the knowledge of the full phase space probability is required in order to compute the correlation, which is clearly strongly dependent by the details of the model. In equilibrium statistical mechanics, a great outcome is that it is possible to recast, under general conditions, the second member of (2.15) in a clear way, as described in Sect. 2.2.1.3. In other words, this is another example of how, at equilibrium, it is possible to get rid of the dynamical details of the model, as it happens for ensemble theory.

In order to fix notation, consider $\mathbf{x}$ as the collection of the phase space variables, then the probability distribution of a trajectory can be written as:

$$\mathcal{P}_0(\omega) = \rho_0(\mathbf{x}) K_0(\omega), \tag{2.16}$$

where ρ_0 is the distribution of the initial conditions. In the following we will present two ways of calculating the response of a generic system: the formal expressions are different, but they are evidently equal, as we will show explicitly at equilibrium. Depending on the model under analysis, it can be convenient to use one expression instead of the other.

2.2.1.1 Linear Response and Steady State Distribution

At the first step we study the behavior of one component of $\mathbf{x}$, say x_i, described by $\rho_{inv}(\mathbf{x})$, which is a non-vanishing and smooth enough invariant measure. When such a system is subjected to an initial perturbation such that $\mathbf{x}(0) \to \mathbf{x}(0) + \Delta\mathbf{x}_0$. We consider the case in which the system is prepared in its steady state, therefore $\rho_0(\mathbf{x}) = \rho_{inv}(\mathbf{x})$. This instantaneous kick modifies the initial density of the system but does not affect the transition rates, therefore one has:

$$\begin{aligned} h &= \Delta\mathbf{x}_0 \\ \rho_h(\mathbf{x}) &= \rho_0(\mathbf{x} - \Delta\mathbf{x}_0) \\ K_h(\omega) &= K_0(\omega). \end{aligned} \tag{2.17}$$

For an infinitesimal perturbation $\delta\mathbf{x}(0) = (0, \ldots, \delta x_j(0), \ldots, 0)$, by substituting (2.17) inside (2.15) one arrives straightforward to[4]

$$R_{i,j}(t) \equiv \frac{\overline{\delta x_i(t)}}{\delta x_j(0)} = -\left\langle x_i(t) \left. \frac{\partial \ln \rho(\mathbf{x})}{\partial x_j} \right|_{t=0} \right\rangle, \tag{2.18}$$

which is the response function of the variable x_i with respect to a perturbation of x_j. This is a first example of a generalized fluctuation response relation, derived for the first time in [32]. The information requested to compute the response in terms of unperturbed correlations is now reduced to the knowledge of the steady state

[4] We put $\rho \equiv \rho_{inv}$ for simplicity.

distribution but can be still non-trivial. However the nature of the perturbation can be easily implemented in numerical experiments: we will describe an application to granular materials in Sect. 2.3.2.

With similar passages, it is also possible to derive the relaxation to finite time perturbation, defining $\overline{\Delta x_i} = \langle x_i \rangle_h - \langle x_i \rangle_0$, from (2.17) and (2.14) one has

$$\overline{\Delta x_i}\,(t) = \Big\langle x_i(t)\; F(\mathbf{x}_0, \Delta\mathbf{x}_0)\Big\rangle, \tag{2.19}$$

where

$$F(\mathbf{x}_0, \Delta\mathbf{x}_0) = \left[\frac{\rho(\mathbf{x}_0 - \Delta\mathbf{x}_0) - \rho(\mathbf{x}_0)}{\rho(\mathbf{x}_0)}\right]. \tag{2.20}$$

In this example, the dependence on the perturbation parameter is highly non linear; this is important in different situations, such as in geophysical or climate investigations: in these contexts, understanding the relaxation to a finite perturbation due to a sudden external change is quite common and represents a challenging issue in comparison to the infinitesimal perturbation required by the linear response theory [9, 57], which can never be applied in practical situations.

2.2.1.2 Linear Response from Transition Rates

In some cases the distribution function is not known and the perturbation enters directly in the equations of motion in form of external field. In these cases a computation from dynamics can be tempted.

Let us define $A(\omega) \equiv -\ln\frac{P_h(\omega)}{P_0(\omega)}$. This functional can be decomposed in two contributions

$$\mathcal{A}(\omega) = \frac{1}{2}(\mathcal{T} - \mathcal{S}), \tag{2.21}$$

where

$$\begin{aligned} \mathcal{T} &= \mathcal{A}(\mathcal{I}\omega) + \mathcal{A}(\omega), \\ \mathcal{S} &= \mathcal{A}(\mathcal{I}\omega) - \mathcal{A}(\omega). \end{aligned} \tag{2.22}$$

The ω dependence in $\mathcal{T}$ and $\mathcal{S}$ is omitted and the time reversal operator $\mathcal{I}$ is introduced. With this formal operation one has:

$$\begin{aligned} \frac{\delta\langle O(t)\rangle_h}{\delta h(t')} &= \frac{\delta\langle O(t)e^{-\mathcal{T}/2+\mathcal{S}/2}\rangle}{\delta h(t')} \\ &= \frac{1}{2}\left\langle O(t)\;\frac{\delta\mathcal{S}}{\delta h(t')}\bigg|_{h=0}\right\rangle - \frac{1}{2}\left\langle O(t)\;\frac{\delta\mathcal{T}}{\delta h(t')}\bigg|\, h=0\right\rangle. \end{aligned} \tag{2.23}$$

It is clear that (2.23) is exactly the same of (2.15), apart from a different notation. In order to go beyond on this result one must restrict to the Markovian case with transition rates $W(x \to y)$ and introduce a prescription for the perturbed transition rates W_h

$$W_h(x \to y) = W(x \to y)e^{\beta/2h(t)[y-x]}. \tag{2.24}$$

Equation (2.24) is called "local detailed balance condition". From this particular assumption, one can derive this expression (see Appendix A.1 for details)

$$R_{Ox}(t,t') = \frac{\beta}{2}[\langle O(t)\dot{x}(t')\rangle - \langle O(t)B(t')\rangle], \tag{2.25}$$

where

$$B(t) \equiv \sum_{y \neq x} W(x \to y)[y - x]. \tag{2.26}$$

Let us stress again that formula (2.25) holds for non-stationary, aging processes, even in absence of detailed balance [3, 22, 63].

At a first sight the two formulas (2.25) and (2.18) appear very different. Actually such a difference can be exploited: we will see in this work how can be convenient one of the two forms with respect to the other, depending on the model under analysis [80].

2.2.1.3 The Equilibrium Case

As mentioned above, linear response theory historically has been developed in an equilibrium context, and many results have been obtained. Let us show how the usual forms of fluctuation dissipation theorems can be deduced from the dynamical versions of the linear response. The advantage of this derivation is that the main features of an equilibrium system must be taken into account and it appears clear how the fluctuation dissipation theorem is a signature of equilibrium.

Let us start from the following identity (see Appendix A.1 for the details of the calculations):

$$\frac{d}{dt}C(t,t') \equiv \langle \dot{x}(t)O(t')\rangle = \langle B(t)O(t')\rangle \quad \text{for } t > t'. \tag{2.27}$$

Moreover, let us consider that:

- if the system is time translational invariant $\frac{d}{dt}C(t,t') = -\frac{d}{dt'}C(t-t')$
- if the system is also invariant for time reversal symmetry $\langle B(t)O(t')\rangle = \langle O(t) B(t')\rangle$.

Within these assumptions from (2.27) and (2.25)

$$R_{Ox}(t) = \beta\langle O(t)\dot{x}(0)\rangle, \tag{2.28}$$

which is nothing but the celebrated fluctuation dissipation theorem. The predictive power of (2.28) is evident especially if compared with the more general relations (2.18) and (2.25): the response of a generic observable is predicted by a suitable correlator, which contains only the observable conjugated to the applied force and is not dependent on the dynamical details of the system under investigation.

It is instructive to recover this result starting also from (2.18). In Hamiltonian systems, taking the canonical ensemble as the equilibrium distribution, one has

$$\ln \rho = -\beta H(\{p\}, \{q\}) + const. \tag{2.29}$$

Then, from the Hamilton's equations ($dq_k/dt = \partial H/\partial p_k$) and from (2.18) one has the differential form of the usual fluctuation dissipation relation [50, 51]:

$$\frac{\overline{\delta O(t)}}{\delta p_k(0)} = \beta \left\langle O(t) \frac{dq_k(0)}{dt} \right\rangle = -\beta \frac{d}{dt} \left\langle O(t) q_k(0) \right\rangle. \tag{2.30}$$

Apart from some differences in the notations, it is evident that (2.28) and (2.30) are completely equivalent.

Moreover, let us suppose to make a perturbation on the momenta p_0. From (2.18), if the distribution of velocities is Maxwell-Boltzmann, one has

$$\frac{\overline{\delta v(t)}}{\delta v(0)} = \beta \langle v(t) v(0) \rangle, \tag{2.31}$$

where, for simplicity of notations, we have introduced the velocity $v \equiv p_0/m$ where m is the mass. Equation (2.31) is most known in its integrated version: if a perturbation like $F\Theta(t)$ acting on the particle is considered,[5] the well known Einstein relation is derived:

$$\mu = \beta D, \tag{2.32}$$

where we have introduced the mobility $\mu \equiv \lim_{t\to\infty} \frac{\overline{\delta v(t)}}{F}$ and the diffusion coefficient

$$D \equiv \int_0^{+\infty} \langle v(t) v(0) \rangle dt. \tag{2.33}$$

We will return largely on (2.32) in Chap. 5, dealing with system exhibiting anomalous diffusion.

In both these derivations, it emerges that when equilibrium dynamics is considered, the response function appears in a compact and general form, involving only the correlation of the observable of interest and the one coupled to the external field. On the contrary, when some currents are flowing into the system and it is driven out of equilibrium, this forms simply fail and no general prescriptions for the response are available. In order to stress this crucial point let us note that, as well clear even

[5] Θ is the Heavyside step function.

in the case of Gaussian variable, the knowledge of a marginal distribution

$$p_i(x_i) = \int \rho(x_1, x_2, ...) \prod_{j \neq i} dx_j \tag{2.34}$$

is not enough for the computation of the autoresponse:

$$R_{i,i}(t) \neq -\left\langle x_i(t) \left. \frac{\partial \ln p_i(x_i)}{\partial x_i} \right|_{t=0} \right\rangle. \tag{2.35}$$

On the contrary, as shown above, the equality in (2.35) holds for the velocities in the case of Maxwell-Boltzmann distribution.

2.2.1.4 The Effective Temperature

In this work we will quite always consider driven systems, with the assumption that they are ergodic and that they reach a steady state in a reasonable time. There is another class of systems where dissipative forces are absent, but they start form an initial configuration which is not the equilibrium one and are, then, characterized by a non-time translational invariant dynamics. In some cases the transient regime has very interesting properties like in domain growth [26], polymers [83], structural glasses [20, 61] and spin glasses [19], where a dramatic slowing down of the relaxation process appears as soon as some parameter is opportunely changed. In these cases the memory of the initial condition is not completely lost and the system "ages": the observables depend non-trivially also by the waiting time, namely the time elapsed since the system is prepared. This "aging regime" is then non stationary and the fluctuation dissipation theorem is not expected to hold; both response and correlation, indeed, decays slower as the system gets older. The analysis of this "fluctuation dissipation violations" has been largely studied in literature (for a review see [24]). In order to give an interpretation of these violations, the concept of effective temperature has been introduced:

$$T^{eff}(t, t_w) = \frac{1}{R(t, t_w)} \frac{\partial C(t, t_w)}{\partial t_w}. \tag{2.36}$$

Note that, up to this moment, (2.36) is just a rewriting and it is obviously true.[6] However, for slow enough dynamics, it is assumed that $T^{eff}(t, t_w) \equiv T^{eff}(C(t, t_w))$, namely the correlation is assumed as a clock of the dynamics. This assumption is verified in mean field glasses. In particular, one can distinguish two well definite regimes: when $(t - t_w)/t_w \ll 1$ one has that $C(t, s) \simeq C_{st}(t - s)$ and the fluctuation dissipation theorem holds, namely $T^{eff}(C)$ is equivalent to the temperature of

[6] A similar definition is often introduced also in the frequency domain conjugated to the variable t [27].

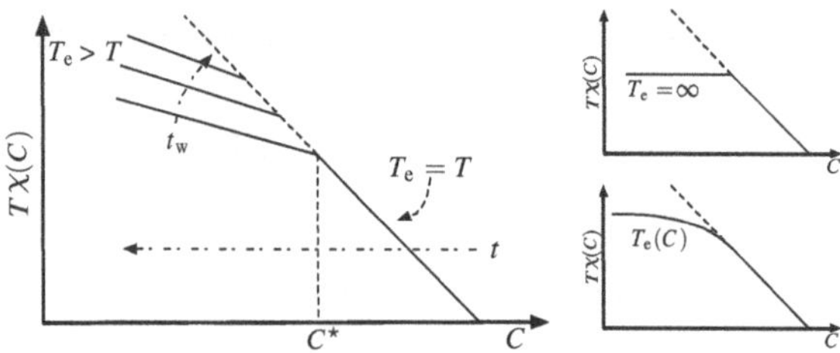

Fig. 2.2 *Left* Integrated response versus correlation in a glass former. For small times (i.e. high values of the correlation) the slope is equal to T and for a value C^* which increases with the waiting time, the slope corresponds to an higher temperature associated to the slow modes. *Right* behavior expected for coarsening (*top*) and spin glasses (*bottom*). The image is taken from [62]

the dynamics T. On the contrary, when $(t - t_w)/t_w > 1$ the fluctuation dissipation theorem is violated and different scenarios are possible, as shown in Fig. 2.2.

It is well established that structural glasses, when quenched below their Mode-Coupling temperature, display an out-of-equilibrium dynamics customarily described within a two-temperature scenario [11, 49, 86]. Fast modes are equilibrated at the bath temperature while slow modes remember, in a sense, the higher temperature determined by the initial condition.

An important question is whether this "violation factor" can be considered a temperature in a thermodynamic sense. This question is still on debate [27, 61], and it has some limits. For instance, it has been shown that, for a generalized version of the trap model of Bouchaud [12], this definition of temperature is observable dependent [35]. Moreover, in some stationary systems like granular gases, the effective temperature meets some conceptual problems, for instance it is negative for vibrated dry granular media [68] and, in case of mixtures, the two components have different temperatures in the steady state [6].

However, it must be noticed that for the class of structural glasses, the interpretation of the fluctuation dissipation ratio (2.36) as an "effective temperature" seems to be well posed, considering also some detailed analysis made on a Leonard-Jones binary mixture showing evidences that the so-defined temperature is observable independent and constant on a long time interval [8].

We will return on the "effective temperature" interpretation on a "two temperature" driven model in Chap. 4.

2.2.2 A Measure of Non-Equilibrium: The Entropy Production

In the previous section we discussed how, when one deals with response theory, it is quite crucial to distinguish between equilibrium and out of equilibrium dynamics. Up to now, we just have stated that non equilibrium regimes are always characterized by some sort of current flowing across the system. Actually, this definition appears quite vague. In this section we go beyond this consideration by introducing a family of observables which somehow can give a sort of distance from equilibrium.

In general, when one deals with non-equilibrium dynamics, very few results, independent from the details of the model, are available. Actually in the last decades a group of relations, known with the name of "fluctuation relations" have captured the interest of the scientific community, especially for their generality and the vast range of applicability. Initially, a numerical evidence given by Evans and Searles [30] showed a particular symmetry in the Cramer function ruling the large deviation of an observable of a molecular fluid under shear. On the other hand, a theorem has been proved by Gallavotti and Cohen [36], under quite general hypothesis, for deterministic systems. This result has been then generalized to stochastic processes by Kurchan [55] and by Lebowitz and Spohn [59]. In a second moment, Jarzinski [45] and Hatano and Sasa [42] have derived other equalities, regarding irreversible transformations: we will return on these last group of identities in Sect. 2.2.3.

Apart from the differences among the various forms of fluctuation relations, it is possible to present these results under an unitary point of view [56], as evidence that the physical ground underlying these results is the same.

According to the description here adopted we will focus on systems in which some noise is present. Thanks to this assumption, it is possible to skip several technical problems and some forms of fluctuation theorems for stochastic systems can be used. We will not enter in the description of the huge literature related to these relations (the interested reader can see, among others [31]) but we will focus on the description of the Lebowitz-Sphon functional, since it is applied to the models presented in the following chapters.

2.2.2.1 The Lebowitz-Sphon Functional

It was shown in Sect. 2.2.1.3 that equilibrium response formula can be derived in a steady state, by assuming time reversal symmetry. This condition is translated on a symmetry property of the probability distribution

$$\mathcal{P}(\omega_t) = \mathcal{P}(\mathcal{I}\omega_t), \tag{2.37}$$

where $\mathcal{I}$ denotes the time reverse operator. Let us consider a transition rate from a generic state x to the state y, from (2.37) the well-known detailed balance condition is obtained

$$\rho_{inv}(x)W(x \to y) = \rho_{inv}(y)W(y \to x). \tag{2.38}$$

where ρ_{inv} is the invariant measure. When a current is present, (2.38) is violated and the time reversal symmetry is broken. From these considerations it appears natural to introduce the following functional for a trajectory of length t:

$$\Sigma_t = \ln \frac{P(\omega_t)}{P(\mathcal{I}\omega_t)}. \tag{2.39}$$

Within this definition, Σ_t is identically equal to zero for each trajectory separately, if detailed balance condition (2.38) is satisfied. Moreover it easy to show, by exploiting the properties of the Kullback-Leibler divergence [53], that $\langle \Sigma_t \rangle$ is always non-negative. Quantity (2.39) is very difficult to be measured, for instance, in an experimental setup [87]. However, in some cases, the entropy production is related to the power injected by external non conservative forces, let us then discuss with a pedagogical example how the entropy production is related to non-equilibrium currents.

Consider a Markov process where the perturbation of an external force F induces non-equilibrium currents. Let us assume that it enters in the transition rates according to local detailed balance condition (2.24), that we rewrite here for clarity

$$\frac{W_F(x \to y)}{W_F(y \to x)} = \frac{W_0(x \to y)}{W_0(y \to x)} e^{2\beta F j(x \to y)}, \tag{2.40}$$

where $j(x \to y)$ is the current associated to the transition $x \to y$, which obey the symmetry property $j(x \to y) = -j(y \to x)$. According to the definition of entropy production (2.39) one finds, for large times,

$$\frac{\Sigma_t}{t} \simeq 2\beta F \frac{1}{t} \sum_{n=1}^{t} j(x(n-1) \to x(n)) = 2\beta F J(t), \tag{2.41}$$

where $J(t)$ is the time-averaged current over a time window of duration t. The fluctuation relation for the probability distribution of the variable $y = \Sigma_t/t$ reads:

$$\frac{P(y)}{P(-y)} = e^y \Longrightarrow \frac{P(2\beta F J(t))}{P(-2\beta F J(t))} = e^{2\beta F J(t)}. \tag{2.42}$$

Namely the fluctuation relation describes a symmetry in probability distribution of the fluctuations of currents. Also, for large times we can assume a large deviation hypothesis $P(y) \sim e^{-tS(y)}$, with $S(y)$ a Cramer function. For small fluctuations around the mean value of y the Cramer function can be approximated to $S(y) = S(2\beta F J) \simeq \beta^2 F^2 (J - \overline{J})^2/\sigma_J^2$, where σ_J is the variance. The fluctuation relation reads as

$$S(y) - S(-y) = y. \tag{2.43}$$

In the Gaussian limit (y close to $\overline{y}$) the previous constraint can be easily demonstrated to be equivalent to $\overline{J}/F = \beta\sigma_J^2$, which is nothing but the standard fluctuation dissipation relation. Therefore the fluctuation relation, which in the simplest case can be directly related to the fluctuation dissipation relation, is a more general symmetry to which we expect to obey the fluctuating entropy production. For a more general discussion of the link between the Lebowitz-Spohn entropy production and currents, see [1, 59]. The remarkable fact appearing in Eq. (2.42) is that it does not contain any free parameter, and so, in this sense, is model-independent.

It is instructive to calculate the entropy production for a simple Langevin equation of a particle in a force field [2]:

$$\dot{v} = -\Gamma v + F(x,t) + \eta(t) \tag{2.44}$$

with, as usual, the noise is Gaussian with $\langle \eta \rangle = 0$, $\langle \eta(t)\eta(t')\rangle = 2T\Gamma\delta(t-t')$, and where $F(x,t) = F_c + F_{nc}$ is a sum of a conservative force $F_c = -U'(x)$ and a non-conservative force $F_{nc}(t)$. The path probability can be written down by introducing the Onsager-Machlup functional [71]:

$$\mathcal{P}(\omega \equiv \{v\}_t) \propto \exp(-L), \tag{2.45}$$

where

$$L = \frac{1}{4\Gamma T}\int_0^t ds\,(\dot{v} + \Gamma v - F)^2 . \tag{2.46}$$

The entropy production reads:

$$\Sigma_t = \ln \frac{P(\omega)}{P(\mathcal{I}\omega)} = \frac{\Delta H}{T} + \frac{\int_0^t F_{nc}(s)v(s)ds}{T}, \tag{2.47}$$

where $\Delta H = \frac{v^2(t)-v^2(0)}{2} + U[x(t)] - U[x(0)]$. Equation (2.47), for large times, allows one to identify the work $w_{nc}(t) = \int_0^t F_{nc}(t)v(t)dt$ done by the external non-conservative force (divided by T) as the entropy produced during the time t. This is an example of the result by Kurchan [54] and by Lebowitz and Spohn [59] about the fluctuation relation for stochastic systems. We will return on this functional in Chap. 3 and 4.

2.2.3 *Entropy Production and the Arrow of Time*

The previous class of fluctuation relations are a sort of extensions of the second law of thermodynamics to small or non-equilibrium systems. In order to see this similarity, let us consider a system $\mathbf{x}$ which moves from the state A to the state B by a variation of a parameter α. Then Hatano and Sasa, showed that

$$\langle e^{-\int dt \frac{\partial \phi(\mathbf{x};\alpha)}{\partial \alpha}\dot{\alpha}} \rangle = 1, \tag{2.48}$$

where $\phi(\mathbf{x};\alpha) = \ln \rho_{inv}(\mathbf{x};\alpha)$, being ρ_{inv} the invariant measure at constant α. By applying the Jensen inequality to (2.48), one has

$$\left\langle -\int dt \frac{\partial \phi(\mathbf{x};\alpha)}{\partial \alpha}\dot{\alpha} \right\rangle \geq 0. \tag{2.49}$$

It is simple to see that the equality is reached only if the transformation is, in a sense, reversible, namely one must assume that, for each value of the control parameter α, the system is in the corresponding stationary state: Eq. (2.48) can be interpreted as a generalization of the second principle of thermodynamic to generic steady states [72]. A relevant question regarding steady states rises again: the quantity $\phi(\mathbf{x};\alpha)$, present in Eq. (2.48) has not a clear thermodynamic meaning, when some currents are present. On the contrary if the system is in equilibrium and the canonical probability density can be assumed, one has

$$\langle e^{-\beta W}\rangle_{A\to B} = e^{-\beta\Delta F}, \tag{2.50}$$

which is named Jarzinski relation, where ΔF is the free energy change between A and B. Also in this case, by means of the Jensen inequality one has that $\langle W\rangle_{A\to B} \geq \Delta F$, which is exactly the second law in thermodynamics. The main message that emerges from this example is that the fluctuation relations of the kind (2.42) are a sort of extension of the second law when fluctuations are relevant. Different connections between these formulas and information theory has been proposed. Let us discuss, for instance, the problem of the arrow of time. In order to fix ideas, let us suppose to observe a trajectory generated by the dynamics of (2.44) and we do not know *apriori* if we are observing it in the right temporal sequence. Clearly if we had an ensemble of trajectories from the same initial condition we could work with the averaged trajectory $\langle v(t)\rangle$ to find easily the answer. On the contrary, because of fluctuations, we cannot be sure of the direction of the time and it becomes a problem of estimation theory. Let be H_0 the hypothesis that the trajectory observed does follow the real time-line and H_1 its negation, a straightforward application of the Bayes formula gives

$$P(H_0|\{V\}) = \frac{P(\{V\}|H_0)P(H_0)}{P(\{V\}|H_0)P(H_0) + P(\{V\}|H_1)P(H_1)}. \tag{2.51}$$

We consider now the case in which there is no reason to prefer as prior an hypothesis respect to the other: then we have $P(H_0) = P(H_1) = \frac{1}{2}$. Moreover $P(\{V\}|H_1) \equiv P(\mathcal{I}\{V\}|H_0)$. Finally, by recalling Eq. (2.47) one has

$$P(H_0|\{V\}) = \frac{1}{1 + e^{-\beta(\Delta H + W_d)}}, \tag{2.52}$$

where the dissipative work has been introduced $W_d = \int_0^t F_{nc}(s)v(s)ds$. Despite of this simplicity, the result (2.52) is really evocative: if the work of the dissipative forces, without sign, is sensibly greater than the thermal fluctuations it is possible to find the correct direction of the time (see Fig. 2.3). Remarkably, if conservative forces are absent, namely if an equilibrium limit is obtained, the probability collapses to the value $\frac{1}{2}$, as a consequence of the detailed balance condition (2.37). This example can by easily generalized to a generic Hamiltonian system showing the same results [10, 46]. Other works have also shown similar connections with the Landauer principle [48].

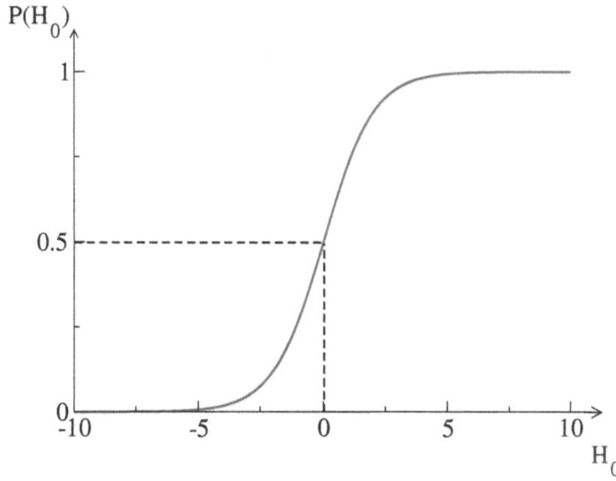

Fig. 2.3 Qualitative behavior of Eq. (2.52) in function of the entropy production, where $\Sigma = \beta(\Delta H + W_d)$. If the dissipative work differs appreciably from zero, it is possible to distinguish the arrow of time of the trajectory. An equilibrium system corresponds to the case $\Sigma = 0$

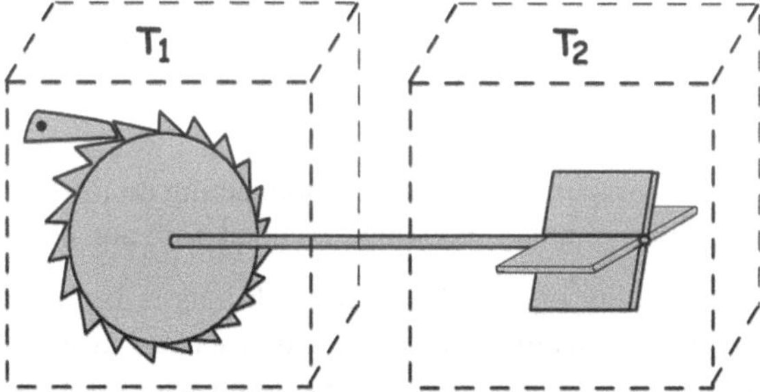

Fig. 2.4 Schematic representation of the Feynman-Smoluchowski ratchet. If the two temperatures are equal, i.e. $T_1 = T_2$, a net drift cannot be observed. The picture is taken from Wikimedia Foundation

These simple arguments are a reflection of the dualism entropy/information. We will return on this subject in Sect. 3.3.

2.2.4 *The Ratchet Effect: A Pure Non-Equilibrium Phenomena*

Let us conclude this part of the chapter by describing a pure non-equilibrium feature, known with the name of ratchet effect. The first one to focus on this problem was Smoluchowski with a *Gedanken experiment* [82], then recovered by Feynman in his popular lectures [34].

As shown in Fig. 2.4, the machine described by Feynman is composed by two compartments. In one of these there is a spring connected with a pawl, while a symmetric rotor is present in the second compartment. Both the compartments are filled with a gas. At a first glance, the machine seems to rotate, since the particles of the gases are supposed to strike uniformly all the faces of the pawl, but it is able to move only in one direction, also if the two temperatures in the compartments are equal. On the contrary, as pointed out by Feynman, the pawl, in order to be sensible to the fluctuation induced by the particles, must be of a similar order of magnitude. Therefore, the dynamics of the pawl is sensible to the thermal fluctuations. Taking into account of this, it is possible to show that there is not a drift. From this example one can conclude that from equilibrium fluctuations is not possible to observe a directed motion. Such a result can be understood in terms of the second law of thermodynamics or, from a kinetic point of view, observing that, from detailed balance condition (2.37), it is not possible to distinguish between past and future. On the contrary, if the two containers of the model are kept at different temperatures ($T_1 \neq T_2$), the system is out of equilibrium and, as commented in Sect. 2.2.2, time reversal symmetry is broken. Under these conditions it is possible to extract work, as derived by Van Den Broeck et al. [15], via kinetic theory, in a simplified version of the model. Note that, in this case, we are not creating work without putting energy into the system: the two reservoirs, indeed, are in contact. Therefore, in a energy balance calculation, also the power injected in order maintain the two temperature different must be taken into account.

As this simple example shows, the necessary ingredient to have a ratchet effect are:

- a spatial symmetry breaking, obtained by an asymmetric shape of intruder or by a non-symmetric external potential acting on the probe particle
- a time symmetry breaking, obtained with non equilibrium conditions.

Even if only one of this two conditions is lacking, a directed motion cannot be observed. This mechanism is also described as a "rectification of non equilibrium fluctuations": let us illustrate this point with a simple overdamped Langevin model [41]:

$$\gamma \dot{x} = -V'(x) + \xi(t), \tag{2.53}$$

where $V(x) \equiv V(x+L)$ is a periodic asymmetric potential with period L and $\xi(t)$ is the usual Gaussian noise with $\langle \xi(t)\xi(t') \rangle = 2\gamma T(t)\delta(t-t')$. Note that the time dependence in the correlation of noise is necessary in order to break the detailed balance condition since the fluctuation dissipation theorem of the second kind is not satisfied. Two typical shapes of the potential and of the temperature are

$$V(x) = V_0 \langle [\sin(2\pi x) + \frac{1}{4}\sin(2\pi x/L)] \rangle \tag{2.54}$$

$$T(t) = T(1 + A\mathrm{sgn}[\sin(\omega t)]). \tag{2.55}$$

where the coefficients in the potential are properly fixed in order to avoid a trivial drift. Within the choice (2.55) the conditions described above are satisfied and a ratchet effect occurs [40]. Given the simplicity of this model, it is not difficult to understand how a proper choice of the shape of $T(t)$ is able to rectify the asymmetries induced by the asymmetric potential, as commented in Fig. 2.5.

In the case above discussed the scales of energy are put by hand and fixed as external parameters, like the two temperatures in Fig. 2.4. On the contrary, there is another class of ratchet whose "second temperature" breaking the detailed balance

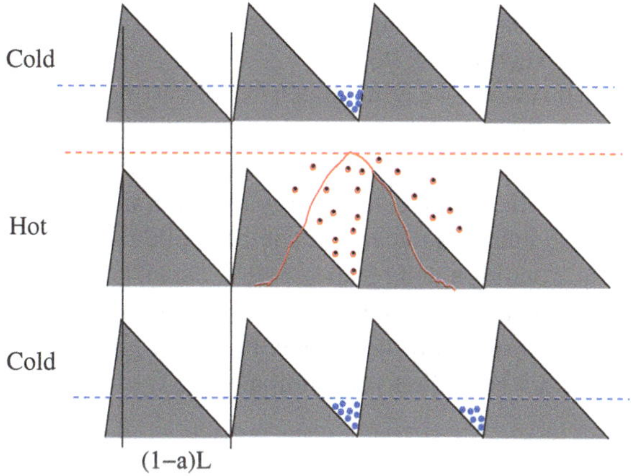

Fig. 2.5 Schematic explanation of a ratchet effect of the kind (2.53). In the hot phase, particle diffuses and, given the asymmetry of the potential, the diffusive motion is rectified in a neat drift. From this example it is clear that the intensity of the drift does non-trivially depend on the other parameters

is, in a sense, generated by non-equilibrium dynamics. The first example in this direction is obtained by substituting the gas with a granular material, characterized by inelastic collisions [23, 65]. Another elegant example has recently experimentally produced by using a thermal bath made of bacteria [60]. In Chap. 4 we will see another application of this kind, by studying an intruder in a fragile glass former. Such an application sounds new for two main reasons: first, it happens in the presence of anomalous transport, and as a consequence, the position of the intruder does not increase linearly in time, secondly the system under investigation is neither stationary nor periodic, as in the other cases here presented.

2.3 An Example of Out of Equilibrium Systems: The Granular Gases

Granular materials are a good candidate to study the non equilibrium effects described up to this point. The main ingredients necessary to have a granular material are essentially the inelastic nature of the collisions and the presence of an excluded volume, due to the macroscopic dimensions of its constituents [47]. Despite of this simplicity there is a huge variety of phenomena that has interested the physicist in the last decades, both in applied and theoretical contexts. It is quite usual to divide granular materials in two main classes: stable or metastable systems and flowing granular systems. One of the first examples of the peculiar properties of the granular material, the quite popular Janssen effect [44], belongs to the first class and describes the deviation from the Stevino's law in such a material. The study of the distribution of avalanches led to introduce the concept of self organized criticality [4]. On the contrary, as the name suggests, in the case of flowing granular systems an uninterrupted flow is present. Also in this regime, several non-equilibrium effects may arise and have been largely studied, like segregation phenomena, pattern formations and convection [37, 43].

In this work, we do not touch the interesting issue of non-ergodic properties related to granular materials, but we always refer to a "granular gas regime", in a steady state condition. In order to reach a steady state it is necessary to balance the dissipation due to collisions with an energy injection mechanism. There are several models of external energy sources that can be applied such that the system rapidly forgets the initial condition and reaches a steady state [75]. We will focus on a specific model, that is the one used in Chaps. 4 and 5, but it must be noticed that, apart from some details, in quite all the models of driven granular gases the scenario described in Sect. 2.3.1 is qualitatively similar.

2.3.1 A Model of a Granular Gas with Thermostat

Let us consider a d-dimensional model for driven granular gases [69, 76, 77, 85]: N identical disks (in $d = 2$) or rods of diameter 1 (in $d = 1$) in a volume $V = L \times L$ or total length L with inelastic hard core interactions characterized by an instantaneous velocity change

$$\mathbf{v}_i' = \mathbf{v}_i - \frac{1+r}{2}[(\mathbf{v}_i - \mathbf{v}_j) \cdot \hat{\sigma}]\hat{\sigma}, \tag{2.56}$$

where i and j are the label of the colliding particles, $\mathbf{v}$ and $\mathbf{v}'$ are the velocity before and after the collision respectively, $\hat{\sigma}$ is the unit vector joining the centers of particles and $r \in [0, 1]$ is the restitution coefficient which is equal to 1 in the elastic case. Each particle i is coupled to a "thermal bath", such that its dynamics (between two successive collisions) obeys

$$m\frac{d\mathbf{v}_i}{dt} = -\frac{1}{\tau_b}\mathbf{v}_i + \sqrt{\frac{2T_b}{\tau_b}}\phi_i(t), \tag{2.57}$$

where τ_b and T_b are parameters of the "bath" and $\phi_i(t)$ are independent normalized white noises. As anticipated before, we restrict ourselves to the dilute or liquid-like regime, excluding more dense systems where the slowness of relaxation prevents clear measures and poses doubts about the stationarity of the regime and its ergodicity.

Note that, with the choice of this kind of thermostat, the equilibrium limit is well defined: if $r = 1$ particles interact each other with elastic collisions and the distribution of velocity is Maxwell-Boltzmann.

Two important observables of the system are the mean free time between collisions τ_c, and the packing fraction ψ. Moreover, it is common to introduce the granular temperature

$$T_g = \frac{m\sum_i \langle v_i^2 \rangle}{N}, \tag{2.58}$$

which is a quite involved function of the parameters, as we will see in the next section.

In this model is possible to recover two different regimes:

- When $\tau_c \gg \tau_b$ grains thermalize, on average, with the bath before experiencing a collision and the inelastic effects are negligible. This is an "equilibrium-like" regime, similar to the elastic case $r = 1$, where the granular gas is spatially homogeneous, the distribution of velocity is Maxwellian and $T_g = T_b$.
- When $\tau_c \ll \tau_b$, non-equilibrium effects can emerge such as deviations from Maxwell-Boltzmann statistics, spatial inhomogeneities and $T_g < T_b$ [69, 76, 77, 85]. This "granular regime", easily reached when packing fraction or inelasticity are increased, is characterized by strong correlations among different particles.

2.3.1.1 Granular Temperature of the Gas

In this section, in order to see an example of a kinetic calculation, we will see how to obtain, in some limit, an expression for the granular temperature T_g.

Multiplying Eq. (2.57) by $\mathbf{v}(t)$ and averaging, one gets[7]

$$\frac{1}{2}m\frac{d}{dt}\langle \mathbf{v}^2(t)\rangle = -\gamma_b\langle \mathbf{v}(t)^2\rangle + \langle \mathbf{v}(t)\mathbf{f}(t)\rangle + \langle \mathbf{v}(t)\eta(t)\rangle. \tag{2.59}$$

Where, for simplicity, we have introduced $\gamma_b = \frac{1}{\tau_b}$ and $\eta_i = \frac{2T_b}{\tau_b}\phi_i$. At stationarity, the left hand side of the above equation vanishes and $\langle \mathbf{v}(t)\eta(t)\rangle = 2\gamma_b T_b/m$. The term $\langle \mathbf{v}(t)\mathbf{f}(t)\rangle$ represents the average power dissipated by collisions:

$$\langle \mathbf{v}(t)\mathbf{f}(t)\rangle = -\langle \Delta E\rangle_{col}, \tag{2.60}$$

where $\Delta E = 1/8m(1-\alpha^2)[(\mathbf{v}_1 - \mathbf{v}_2)\cdot\hat{\mathbf{e}}]^2$ is the energy dissipated per particle and the collision average is defined by

$$\langle \ldots\rangle_{col} = \int d\hat{\mathbf{e}}\int d\mathbf{v}_1\int d\mathbf{v}_2 \ldots p(\mathbf{v}_1,\mathbf{v}_2)\Theta[-(\mathbf{v}_1-\mathbf{v}_2)\cdot\hat{\mathbf{e}}]|(\mathbf{v}_1-\mathbf{v}_2)\cdot\hat{\mathbf{e}}|.$$

This integral contains the joint distribution of the collisional particle velocity. It can be solved with the Enskog correction, a slight modification of the molecular chaos assumption [21]:

$$p(\mathbf{v}_1,\mathbf{v}_2) = \chi p(\mathbf{v}_1)p(\mathbf{v}_2), \tag{2.61}$$

where $\chi = \frac{g_2'(2r)}{l_0}$ and l_0 is the mean free path and $g_2'(2r)$ is the pair correlation function for two gas particles at contact. Equation (2.61) is expected to hold in a dilute system, but fails in denser regimes, because of recollisions and memory effects.

Thanks to this approximation, the integral in Eq. (2.60) can be computed by standard methods [13], and, in two dimensions within the Gaussian approximation, yields

[7] The index i has been removed for simplicity.

$$\langle \Delta E \rangle_{col} = \chi_g \frac{\sqrt{\pi}(1-r^2)}{\sqrt{m}} T_g^{3/2}. \tag{2.62}$$

Substituting this result into Eq. (2.59) and recalling that $T_g = m\langle \mathbf{v}^2 \rangle /2$, one finally obtains the implicit equation

$$T_g = T_b - \chi_g \frac{\sqrt{\pi m}(1-r^2)}{2\gamma_b} T_g^{3/2}, \tag{2.63}$$

which can be solved to obtain T_g. Note that, from (2.63), when $\gamma_b \to \infty$, the equilibrium-like limit is recovered and $T_b = T_g$.

2.3.2 Response Analysis

For the model presented above, and for other similar steady state granular gases, a response analysis has been performed [5, 7, 14, 74, 78, 84].

We will focus on the numerical experiments on the model described in Sect. 2.3.1. The protocol used in numerical experiments cited above is the following:

1. The gas is prepared in a "thermal" state, with random velocity components extracted from a Gaussian with zero average and given variance, and positions of the particles chosen uniformly random in the box, avoiding overlapping configurations.
2. The system is let evolve until a statistically stationary state is reached, which is set as time 0.
3. A copy of the system is obtained, identical to the original but for one particle, whose x (for instance) velocity component is incremented of a fixed amount $\delta v(0)$.
4. Both systems are let evolve with the unperturbed dynamics. For the random thermostats, the same noise realization is used. The perturbed tracer has velocity $v'(t)$, while the unperturbed one has velocity $v(t)$, so that $\delta v(t) = v'(t) - v(t)$.
5. After a time t_{max} large enough to have lost memory of the configuration at time 0, a new copy is done with perturbing a new random particle and the new response is measured. This procedure is repeated until a sufficient collection of data is obtained.
6. Finally the autocorrelation function $C_{vv}(t) = \langle v(t)v(0)\rangle$ in the original system and the response $R_{vv}(t) \equiv \overline{\frac{\delta v(t)}{\delta v(0)}}$ are measured.

In dilute cases, it is numerically observed that the phase space distribution can be factorized, namely:

$$\rho(\{\mathbf{v}_i, x_i\}) = n^N \prod_{i=1}^{N} \prod_{\alpha=1}^{d} p_v(v_i^{(\alpha)}), \tag{2.64}$$

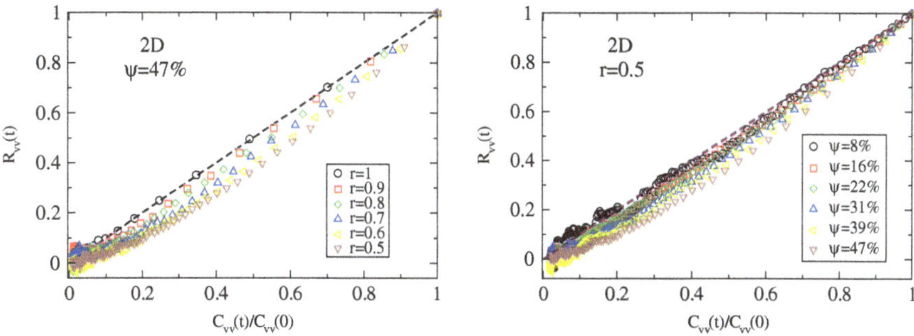

Fig. 2.6 Parametric plots to check the Einstein Relation, for $d = 2$ models of inelastic hard-core gases with thermal bath. Different choices of parameters r (restitution coefficient), $\alpha = \tau_c/\tau_b$ and ψ (packing fraction) are shown: note that one can change α at ψ or r fixed (changing τ_b), but - in general - changes in ψ or r determine also changes in α (because of changes in τ_c). In all plots, the dashed line marks the Einstein relation $R_{vv} = C_{vv}(t)/C_{vv}(0)$

with n the spatial density $n = N/V$ and $p_v(v)$ the one-particle velocity component probability density function, $v_i^{(\alpha)}$ the α-th component of the velocity of the i-th particle and d the system dimensionality. Exploiting isotropy, we will denote with v an arbitrary component of the velocity vector: the results do not change if v is the x or y component.

From (2.18), it is expected that an instantaneous perturbation $\delta v(0)$, at time $t = 0$ on a particle of the gas will result in an average response of the form

$$R(t) = \frac{\overline{\delta v(t)}}{\delta v(0)} = -\left\langle v(t) \frac{\partial \ln p_v(v)}{\partial v}\bigg|_0 \right\rangle \neq C_1(t), \tag{2.65}$$

having defined $C_1(t) = \langle v(t)v(0)\rangle / \langle v^2 \rangle$. On the contrary it is observed that noticeable deviations form Einstein relation do not occur, therefore non-Gaussianity alone is not sufficient to produce violations. Indeed it has been shown in simplified models that all the higher order correlations are proportional [74]

$$C_f(t) = \frac{\langle v(t) f[v(0)] \rangle}{\langle v(0) f[v(0)] \rangle} \approx C_1(t), \tag{2.66}$$

which is shown to be valid also in the model here described, by numerical inspection. In conclusion, in the dilute limit the two conditions (2.64) and (2.66) are sufficient to verify the Einstein relation.

On the contrary, when the system is denser, the Molecular chaos approximation is no more valid and Eq. (2.64) fails; as a consequence, one can observe strong deviations from linearity between response and autocorrelation.

In addiction, there are same remarkable points: the violation is more and more pronounced as the inelasticity increases (lower values of r), the importance of the bath is reduced (lower values of τ_b/τ_c) or the packing fraction is increased, as shown in

Fig. 2.6. In correspondence of such variations of parameters, the correlation between velocities of adjacent particles is also enhanced, a phenomena which is ruled out in equilibrium fluids. We will return on this aspect in Chap. 5 by observing a similar behavior in a one dimensional model.

2.3.3 Entropy Production in Granular Gases: A Challenge

An experiment has been performed by Menon and Feitosa [33] using a granular gas shaken in a container at high frequency. The setup consisted of a $2D$ vertical box containing N identical glass beads, vertically vibrated at frequency f and amplitude A. The authors observed the kinetic energy variations ΔE_τ, over time windows of duration τ, in a central sub-region of the system characterized by an almost homogeneous temperature and density. They subdivided this variation into two contributions:

$$\Delta E_\tau = W_\tau - D_\tau, \tag{2.67}$$

where D_τ is the energy dissipated in inelastic collisions and W_τ is the energy flux through the boundaries, due to the kinetic energy transported by incoming and outgoing particles. The authors of the experiment have conjectured that W_τ, being a measure of injected power in the sub-system, can be related to the entropy flow or the entropy produced by the thermostat constituted by the rest of the gas (which is equal to the internal entropy production in the steady state). They have measured its probability distribution $f(W_\tau)$ and found that

$$\ln \frac{f(W_\tau)}{f(-W_\tau)} = \beta W_\tau, \tag{2.68}$$

with $\beta \neq 1/T_g$. By lack of a reasonable explanation for the value of β, the authors have concluded to have experimentally verified the fluctuation relation with an "effective temperature" $T_{eff} = 1/\beta$, suggesting its use as a possible non-equilibrium generalization of the usual granular temperature. The same results have been found in molecular dynamics simulations of inelastic hard disks with a similar setup [79]. This "effective temperature" interpretation is not convincing for different reasons. Among others, in the elastic limit one should expect that this definition of temperature should coincide with the external one, on the contrary it diverges, since the function $f(W_\tau)$ is symmetric. A different explanation has been proposed in [79], and it shows no connection with the fluctuation relation. It appears that the injected power measured in the experiment can be written as

$$W_\tau = \frac{1}{2}\left(\sum_{i=1}^{n_+} v_{i+}^2 - \sum_{i=1}^{n_-} v_{i-}^2\right), \tag{2.69}$$

where n_- (n_+) is the number of particles leaving (entering) the sub-region during the interval of time τ. In this case the authors assume n_- and n_+ being Poisson-distributed, neglecting correlations among particles entering or leaving successively the central region. The key ingredient due to inelasticity is that, as confirmed by simulations, the velocities $\mathbf{v}_{i+}$ and $\mathbf{v}_{i-}$ are assumed to originate from two distinct populations with different temperatures T_+ and T_- respectively. Within this assumptions, the left member of (2.68) can be exactly calculated, showing a non linear behavior in W_τ. The linear expression (2.68) found in the experiments is consistent with the expansion up to the third order[8] and can then be explained with lack of statistics.

2.3.4 Some Remarks

As underlined in the previous sections, both the response properties and the entropy production in granular gases are non completely clarified issues.

Regarding response properties, the lack of factorization of the phase space distribution produces a failure of the Einstein relation. A tentative modelling of the response was tried in [74] by assuming an effective distribution for the perturbed particle

$$p_v(v, \mathbf{x}, t) \sim \exp\left\{-\frac{(v - u(\mathbf{x}, t))^2}{2T_g}\right\}, \tag{2.70}$$

where $u(\mathbf{x}, t)$ is an effective fluctuating local velocity field coupled to the tracer (one can think to the average velocity of the surrounding particles). Using (2.18) one has

$$\frac{\overline{\delta v(t)}}{\delta v(0)} \propto \langle v(t)\,[v(0) - u(\mathbf{x}, t)]\rangle. \tag{2.71}$$

This formula has the merit to catch the main non-equilibrium source of the syste; for this reason, indeed, there is a partial agreement with simulations. On the other hand, assumption (2.70) has two main problems. First of all, a local velocity field can be defined only with an associated length, and, at this level, there is not an operative definition of it. Second, for the elastic limit the Einstein relation must be recovered and the distribution of velocities must approach to a Maxwell-Boltzmann; such a limit is not straightforward in Eq. (2.70), since a local velocity field does also exist in a dense elastic case, but is uncoupled to the tracer. Noticeably, one of the major criticism that can be moved to the "effective temperature interpretation" (2.68) is that the elastic limit is not well defined.

Given the considerations above, it appear necessary to work in a more controlled setup, where the correlations and the corresponding coupling with a velocity field emerge as soon as dissipation due to inelasticity is turned on, and a proper elastic limit can be recovered, together with the validity of the fluctuation dissipation theorem

[8] The second order term trivially vanishes for functions of the form $g(x) = \ln[f(x)/f(-x)]$.

and with a vanishing entropy production. This issue is a central point in this work and it will be discussed in Chaps. 3 and 4.

References

1. Andrieux, D., Gaspard, P.: Fluctuation theorem for currents and schnakenberg network theory. J. Stat. Phys. **127**, 107 (2007)
2. Astumian, R.D.: The unreasonable effectiveness of equilibrium theory for interpreting non-equilibrium experiments. Am. J. Phys **74**, 683 (2006)
3. Baiesi, M., Maes, C., Wynants, B.: Nonequilibrium linear response for markov dynamics, i: jump processes and overdamped diffusions. J. Stat. Phys. **137**, 1094 (2009)
4. Bak, P., Tang, C., Wiesenfeld, K.: Self-organized criticality. Phys. Rev. A **38**, 364 (1988)
5. Baldassarri, A., Barrat, A., D'Anna, G., Loreto, V., Mayor, P., Puglisi, A.: What is the temperature of a granular medium? J. Phys. Condens. Matter **17**, S2405 (2005)
6. Barrat, A., Trizac, E.: Lack of energy equipartition in homogeneous heated binary granular mixtures. Granular Matter **4**, 57 (2002)
7. Barrat, A., Loreto, V., Puglisi, A.: Temperature probes in binary granular gases. Phys. A. **334**, 513 (2004)
8. Berthier, L., Barrat, J.L.: Shearing a glassy material: numerical tests of nonequilibrium mode-coupling approaches and experimental proposals. Phys. Rev. Lett. **89**, 095702 (2002)
9. Boffetta, G., Lacorata, G., Musacchio, S., Vulpiani, A.: Relaxation of finite perturbations: Beyond the fluctuation-response relation. Chaos **13**, 806 (2003)
10. Boksenbojm, E., Wynants, B., Jarzynski, C.: Nonequilibrium thermodynamics at the microscale: work relations and the second law. Stat. Mech. Appl. Phys A **389**, 4406 (2010)
11. Bouchaud, J.P., Cugliandolo, L.F., Kurchan, J., Mezard. M.: Spin Glasses and Random Fields. World Scientific, singapore (1998)
12. Bouchaud, J., Dean, D.: Aging on parisi's tree. J. Phy. **I**(5), 265 (1995)
13. Brilliantov, N.K., Poschel, T.: Kinetic Theory of Granular Gases. Oxford University Press, Oxford (2004)
14. Brilliantov, N.V., Poschel, T.: Self-diffusion in granular gases: green-kubo versus chapman-enskog. Chaos **15**, 026108 (2005)
15. Van den Broeck, C., Kawai, R., Meurs, P.: Microscopic analysis of a thermal brownian motor. Phys. Rev. Lett. **93**, 90601 (2004)
16. Brown, R.: A brief account of microscopical observations made.. on the particles contained in the pollen of plants; and on the general existence of active molecules in organic and inorganic bodies. Phil. Mag. Ser. 2 **4**, 161 (1828)
17. Brown, R.: Additional remarks on active molecules. Phil. Mag Ser. 2 **6**, 161 (1829)
18. Brown's microscopical observations on the particles of bodies. Philos. Mag. N. S., **8**, 296 (1830)
19. Castellani, T., Cavagna, A.: Spin-glass theory for pedestrians. J. Stat. Mech. Theor. Exp. **2005**, P05012 (2005)
20. Cavagna, A.: Supercooled liquids for pedestrians. Phys. Rep. **476**, 51 (2009)
21. Chapman, S., Cowling, T.: The mathematical theory of non-uniform gases. The mathematical theory of non-uniform gases, vol. 1. Cambridge University Press, Cambridge (1991).
22. Corberi, F., Lippiello, E., Sarracino, A., Zannetti, M.: Fluctuation-dissipation relations and field-free algorithms for the computation of response functions. Phys. Rev. E **81**, 011124 (2010)
23. Costantini, G., Marconi, U., Puglisi, A.: Granular brownian ratchet model. Phys. Rev. E **75**, 061124 (2007)
24. Crisanti, A., Ritort, F.: Violation of the fluctuation-dissipation theorem in glassy systems: basic notions and the numerical evidence. J. Phys. A **36**, R181 (2003)

25. Cugliandolo, L.: Disordered systems. Lecture notes, Cargese (2011)
26. Cugliandolo, L.F.: Weak-ergodicity breaking in mean-field spin-glass models. Phil. Mag. **71**, 501–514 (1995)
27. Cugliandolo, L., Kurchan, J., Peliti, L.: Energy flow, partial equilibration, and effective temperatures in systems with slow dynamics. Phys. Rev. E **55**, 3898 (1997)
28. Dunkel, J., Hänggi, P.: Relativistic brownian motion. Phys. Rep. **471**, 1 (2009)
29. Einstein, A.: On the movement of small particles suspended in a stationary liquid demanded by the molecular-kinetic theory of heat. Ann. d. Phys. **17**, 549 (1905)
30. Evans, D.J., Searles, D.J.: Equilibrium microstates which generate second law violating steady states. Phys. Rev. E **50**, 1645 (1994)
31. Evans, D.J., Searles, D.J.: The fluctuation theorem. Adv. Phys. **52**, 1529 (2002)
32. Falcioni, M., Isola, S., Vulpiani, A.: Correlation functions and relaxation properties in chaotic dynamics and statistical mechanics. Phys. Lett. A **144**, 341 (1990)
33. Feitosa, K., Menon, N.: Fluidized granular medium as an instance of the fluctuation theorem. Phys. Rev. Lett. **92**, 164301 (2004)
34. Feynman, R., Leighton, R., Sands, M., et al.: The Feynman lectures on physics, vol. 2. Addison-Wesley Reading, MA (1964)
35. Fielding, S., Sollich, P.: Observable dependence of fluctuation-dissipation relations and effective temperatures. Phys. Rev. Lett. **88**, 50603 (2002)
36. Gallavotti, G., Cohen, E.G.D.: Dynamical ensembles in stationary states. J. Stat. Phys. **80**, 931 (1995)
37. Goldhirsch, I.: Rapid granular flows. Ann. Rev. Fluid Mech. **35**, 267 (2003)
38. Hänggi, P.: Generalized langevin equations: a useful tool for the perplexed modeller of non-equilibrium fluctuations? Stochastic, dynamics, p. 15. Springer, Berlin (1997)
39. Hänggi, P., Ingold, G.: Fundamental aspects of quantum brownian motion. Chaos: an Interdisciplinary. J. Nonlinear Sci. **15**, 026105 (2005)
40. Hänggi, P., Marchesoni, F., Nori, F.: Brownian motors. Ann. Phys. **14**, 51 (2005)
41. Hänggi, P., Marchesoni, F.: Artificial brownian motors: controlling transport on the nanoscale. Rev. Mod. Phys. **81**, 387 (2009)
42. Hatano, T., Sasa, S.: Steady-state thermodynamics of Langevin systems. Phys. Rev. Lett. **86**, 3463 (2001)
43. Jaeger, H.M., Nagel, S.R.: Physics of the granular state. Science **255**, 1523 (1992)
44. Janssen, H. Versuche uber getreidedruck in silozellen. z. ver deut. Ing. **39** 1045 (1895)
45. Jarzynski, C.: Nonequilibrium equality for free energy differences. Phys. Rev. Lett. **78**, 2690 (1997)
46. Jarzynski, C.: Nonequilibrium work relations: foundations and applications. Eur. Phys. J. B Condens. Matter Complex Syst. **64**, 331 (2008)
47. Kadanoff, L.P.: Built upon sand: theoretical ideas inspired by granular flows. Rev. Mod. Phys. **71**, 435 (1999)
48. Kawai, R., Parrondo, J., den Broeck, C.: Dissipation: the phase-space perspective. Phys. Rev. Lett. **98**, 80602 (2007)
49. Kob, W., Barrat, J., Sciortino, F., Tartaglia, P.: Aging in a simple glass former. J. Phys. Condens. Matter **12**, 6385 (2000)
50. Kubo, R.: The fluctuation-dissipation theorem. Rep. Prog. Phys. **29**, 255 (1966)
51. Kubo, R.: Brownian motion and nonequilibrium statistical mechanics. Science **32**, 2022 (1986)
52. Kubo, R., Toda, M., Hashitsume, N.: Statistical physics II Nonequilibrium Stastical Mechanics. Springer, Berlin (1991)
53. Kullback, S., Leibler, R.: On information and sufficiency. Ann. Math. Stat. **22**, 79 (1951)
54. Kumaran, V.: Temperature of a granular material "fluidized" by external vibrations. Phys. Rev. E **57**, 5660 (1998)
55. Kurchan, J.: Fluctuation theorem for stochastic dynamics. J. Phys. A **31**, 3719 (1998)
56. Kurchan, J.: Non-equilibrium work relations. J. Stat. Mech. Theor. Exp. **2007**, P07005 (2007)
57. Lacorata, G., Puglisi, A., Vulpiani, A.: On the fluctuation-response relation in geophysical systems. Int. J. Mod. Phys. B **23**, 5515 (2009)

58. Langevin, P.: Sur la theorie du mouvement brownien. C. R. Acad. Sci. (Paris) **146** 530 (1908) (Translated in. Am. J. Phys. **65**, 1079 (1997))
59. Lebowitz, J.L., Spohn, H.: A Gallavotti-Cohen-type symmetry in the large deviation functional for stochastic dynamics. J. Stat. Phys. **95**, 333 (1999)
60. Di Leonardo, R., et al.: Bacterial ratchet motors. Proc. Nat. Acad. Sci. **107**, 9541 (2010)
61. Leuzzi, L., Nieuwenhuizen, T.M.: Thermodynamics of the Glassy State. Taylor & Francis, New York (2007)
62. Leuzzi, L.: A stroll among effective temperatures in aging systems: limits and perspectives. J. Non-Cryst. Solids **355**, 686 (2009)
63. Lippiello, E., Corberi, F., Zannetti, M.: Off-equilibrium generalization of the fluctuation dissipation theorem for Ising spins and measurement of the linear response function. Phys. Rev. E **71**, 036104 (2005)
64. Marconi, U., Puglisi, A., Rondoni, L., Vulpiani, A.: Fluctuation-dissipation: response theory in statistical physics. Phys. Rep. **461**, 111 (2008)
65. Van Der Meer, D., Reimann, P., Van Der Weele, K., Lohse, D.: Spontaneous ratchet effect in a granular gas. Phys. Rev. Lett. **92**, 184301 (2004)
66. Mori, H.: Transport, collective motion, and brownian motion. Progress Theoret. Phys. **33**, 423 (1965)
67. Nelson, E. Dynamical theories of Brownian motion. Citeseer **17** (1967)
68. Nicodemi, M.: Dynamical response functions in models of vibrated granular media. Phys. Rev. Lett. **82**, 3734 (1999)
69. van Noije, T.P.C., Ernst, M.H., Trizac, E., Pagonabarraga, I.: Randomly driven granular fluids: large-scale structure. Phys. Rev. E **59**, 4326 (1999)
70. Onsager, L.: Reciprocal relations in irreversible processes. I. Phys. Rev. **37**, 405 (1931)
71. Onsager, L., Machlup, S.: Fluctuations and irreversible processes. Phys. Rev. **91**, 1505 (1953)
72. Perez-Espigares, C., Kolton, A. B., Kurchan, J.: An infinite family of second law-like inequalities. Phys. Rev. **85**(3), 031135 (2012)
73. Perrin, J.: Les Atomes. Alcan, Paris (1913)
74. Puglisi, A., Baldassarri, A., Vulpiani, A.: Violations of the einstein relation in granular fluids: the role of correlations. J. Stat. Mech. P08016 (2007)
75. Puglisi, A.: Granular fluids, a short walkthrough (2010)
76. Puglisi, A., Loreto, V., Marconi, U.M.B., Petri, A., Vulpiani, A.: Clustering and non-gaussian behavior in granular matter. Phys. Rev. Lett. **81**, 3848 (1998)
77. Puglisi, A., Loreto, V., Marconi, U.M.B., Vulpiani, A.: A kinetic approach to granular gases. Phys. Rev. E **59**, 5582 (1999)
78. Puglisi, A., Baldassarri, A., Loreto, V.: Fluctuation-dissipation relations in driven granular gases. Phys. Rev. E **66**, 061305 (2002)
79. Puglisi, A., Visco, P., Barrat, A., Trizac, E., van Wijland, F.: Fluctuations of internal energy flow in a vibrated granular gas. Phys. Rev. Lett. **95**, 110202 (2005)
80. Seifert, U., Speck, T.: Fluctuation-dissipation theorem in nonequilibrium steady states. EPL (Europhys. Lett.) **89**, 10007 (2010)
81. Smoluchowski, M.: Zur kinetischen theorie der brownschen molekularbewegung und der suspensionen. Ann. d. Phys. **21**, 756 (1906)
82. Smoluchowski, M.: Experimentell nachweisbare, der üblichen thermodynamik widersprechende molekularphänomene. Physik. Zeitschr **13**, 1069 (1912)
83. Struik, L.: Physical Aging in Amorphous Polymers and Other Materials. Elsevier, Amsterdam (1978)
84. Villamaina, D., Puglisi, A., Vulpiani, A.: The fluctuation-dissipation relation in sub-diffusive systems: the case of granular single-file diffusion. J. Stat. Mech. L10001 (2008)
85. Williams, D.R.M., MacKintosh, F.C.: Driven granular media in one dimension: correlations and equation of state. Phys. Rev. E **54**, R9 (1996)
86. Zamponi, F., Bonetto, F., Cugliandolo, L. F., Kurchan, J.: A fluctuation theorem for non-equilibrium relaxational systems driven by external forces. J. Stat. Mech. P09013 (2005)

87. Zamponi, F.: Is it possible to experimentally verify the fluctuation relation? a review of theoretical motivations and numerical evidence. J. Stat. Mech. P02008 (2007)
88. Zwangzig, R.: Nonequilibrium statistical mechanics. Oxford University Press, Oxford (2001)
89. Zwanzig, R.: Time-correlation functions and transport coefficients in statistical mechanics. Ann. Rev. Phys. Chem. **16**, 67 (1965)

Chapter 3
The Effects of Memory on Linear Response and Entropy Production

Abstract In this chapter the Langevin equation with memory is analyzed in both equilibrium and non equilibrium setups. Non-Markovian equation can be mapped in a Markovian one by increasing enough the number of degrees of freedom. This procedure is not just a simple mathematical trick, on the contrary the relative coupling between different variables is relevant for the correct prediction of the response. Moreover such a coupling is proportional to the entropy production rate. By concluding, we show how the mapping from Markovian to non-Markovian dynamics is equivalent to a projection operation and it carries a loss of information that can be detected by entropy production.

In what follows we will apply some concepts presented in the previous chapter, focusing on multidimensional Langevin processes. In the first part we propose a derivation of the entropy production for this kind of dynamical equations. The presence of reversible and irreversible probability currents allows one to distinguish, to a more abstract level, between equilibrium and non-equilibrium.

The second part presents an application of entropy production to Generalized Langevin equations, namely Langevin equations with a memory kernel and colored noise. Clearly, memory can emerge also at equilibrium, but when the fluctuation dissipation theorem of second kind is not satisfied, the detailed balance is violated and one obtains pure non-equilibrium dynamics. By introducing auxiliary variables, it is possible to recast the non-Markovian process into a Markovian one and the entropy production rate can be easily calculated, showing the presence of "memory forces". In this sense memory effects, when combined with detailed balance violations, can act as non-equilibrium sources. We analyze in detail the linear model, which has the advantage of being completely analytically treatable but still non-trivial: its analysis reveals the intimate connection between auxiliary variables, correlations, response properties and entropy production.

The mapping from memory to auxiliary variables seems, at a first sight, an innocent reparametrization of a system, without physical consequences. On the contrary, in the last part of this chapter, it is shown that the connection between the two descriptions is given by a projection procedure, which is necessarily connected with a loss of

D. Villamaina, *Transport Properties in Non-Equilibrium and Anomalous Systems*, Springer Theses, DOI: 10.1007/978-3-319-01772-3_3,

information. This reduction can be crucial in some situations; for instance it produces a vanishing entropy production in the linear case.

3.1 Entropy Production in Langevin Processes

Let us first discuss a non-multiplicative multivariate Langevin equation with N degrees of freedom:

$$\dot{X}_i = D_i(\mathbf{X}) + \xi_i(t), \tag{3.1}$$

with $i \in [0, N-1]$, $\xi_i(t)$ is a Gaussian process with $\langle \xi_i(t) \rangle = 0$ and $\langle \xi_i(t)\xi_j(t') \rangle = 2D_{ij}\delta(t-t')$. where D_{ij} is symmetric by construction.

Variables X_i are assumed to have a well-defined parity $\epsilon_i = \pm 1$, with respect to time-reversal. This leads to recognize reversible and irreversible parts of the drift:

$$D_i(\mathbf{X}) = D_i^{rev}(\mathbf{X}) + D_i^{ir}(\mathbf{X}) \tag{3.2}$$

with

$$D_i^{rev}(\mathbf{X}) = \frac{1}{2}[D_i(\mathbf{X}) - \epsilon_i D_i(\epsilon\mathbf{X})] = -\epsilon_i D_i^{rev}(\epsilon\mathbf{X}) \tag{3.3}$$

$$D_i^{ir}(\mathbf{X}) = \frac{1}{2}[D_i(\mathbf{X}) + \epsilon_i D_i(\epsilon\mathbf{X})] = \epsilon_i D_i^{ir}(\epsilon\mathbf{X}) \tag{3.4}$$

having defined $\epsilon\mathbf{X} = (\epsilon_0 X_0, \epsilon_1 X_1, ...\epsilon_{N-1} X_{N-1})$. Following the Onsager-Machlup recipe [13, 16], the expression for conditional path probability of trajectory $\{\mathbf{X}(s)\}_0^t$, is[1]

$$\log P(\{\mathbf{X}(s)\}_0^t) = -\frac{1}{4}\sum_{jk}\int_0^t ds D_{jk}^{-1}\{\dot{X}_j(s) - D_j[\mathbf{X}(s)]\} \times \{\dot{X}_k(s) - D_k[\mathbf{X}(s)]\}, \tag{3.5}$$

where we have assumed that D_{ij}^{-1} exists.

By using (3.2) and a few passages we get

$$\begin{aligned} W_t = &\log\frac{P(\{\mathbf{X}(s)\}_0^t)}{P(\{\mathcal{I}\mathbf{X}(s)\}_0^t)} = -\frac{1}{2}\int_0^t ds D_{jk}^{-1} \\ &\times \left\{a_{jk}^{-}\left[\dot{X}_j\dot{X}_k + D_j^{ir}D_k^{ir} + D_j^{rev}D_k^{rev} - 2\dot{X}_j D_k^{rev}\right]\right. \\ &\left.-2a_{jk}^{+}\left[D_j^{ir}\dot{X}_k - D_j^{ir}D_k^{rev}\right]\right\}, \end{aligned} \tag{3.6}$$

[1] We use the Ito convention for stochastic integrals. Note that, with this convention, the Jacobian in the path probability is 1 [8].

where we have introduced the following definitions:

$$a^{-}_{jk} = \frac{1 - \epsilon_j \epsilon_k}{2}, \qquad a^{+}_{jk} = \frac{1 + \epsilon_j \epsilon_k}{2}. \tag{3.7}$$

Equation 3.6 is strongly simplified in the case of a diagonal diffusion matrix D_{ij}, obtaining:

$$W_t = \sum_k D_{kk}^{-1} \int_0^t ds D_k^{ir} \left[\dot{X}_k - D_k^{rev} \right]. \tag{3.8}$$

3.1.1 The Fluctuation Relation and the Border Terms

The probability distribution $f_t(\mathbf{X})$ of the process satisfies the Fokker–Planck equation

$$\frac{\partial f_t(\mathbf{X})}{\partial t} = -\sum_i \frac{\partial S_i(\mathbf{X})}{\partial X_i}, \tag{3.9}$$

with the probability current defined by

$$S_i(\mathbf{X}) = D_i(\mathbf{X}) f_t(\mathbf{X}) - \sum_j \frac{\partial}{\partial X_j} D_{ij} f_t(\mathbf{X}). \tag{3.10}$$

The decomposition (3.2) can be extended to the probability current:

$$S_i^{rev} = f_t D_i^{rev} \qquad S_i^{ir} = S_i - S_i^{rev} = f_t D_i^{ir} - \sum_j D_{ij} \frac{\partial f_t}{\partial X_j}. \tag{3.11}$$

Equations (3.3) and (3.4) are coherent with the fact that, upon time reversal, $\partial/\partial t$ changes sign too. It can be verified (see [16]) that a necessary and sufficient condition for detailed balance is $S_i^{ir} = 0$ in the stationary state. Therefore, if $S_i^{ir} = 0$, a path and its time-reversal have the same stationary probability; this is not true if $S_i^{ir} \neq 0$.

We now recall that, in order to obtain the complete path probability in the steady state, one has to multiply $P(\{\mathbf{X}(s)\}_0^t)$ by $f(\mathbf{X}(0))$, where $f = \lim_{t \to +\infty} f_t$ is the stationary probability distribution. It is therefore possible to compute a different quantity

$$W_t' = \log \frac{f[\mathbf{X}(0)] P(\{\mathbf{X}(s)\}_0^t)}{f[\epsilon \mathbf{X}(t)] P(\{\mathcal{I}\mathbf{X}(s)\}_0^t)} = W_t + b_t \tag{3.12}$$

$$b_t = \log\{f[\mathbf{X}(0)]\} - \log\{f[\epsilon \mathbf{X}(t)]\}. \tag{3.13}$$

The term b_t is the "border term" [20]. The condition of detailed balance is equivalent to $W' \equiv 0$ for all trajectories in the steady state. When detailed balance does not hold, it can be shown that (see Sect. 2.2.2.1)

$$\log \frac{\mathrm{p}(W_t' = x)}{\mathrm{p}(W_t' = -x)} = x, \tag{3.14}$$

where $p(W_t' = x)$ is the probability in the steady state. Equation (3.14) is the finite-time fluctuation relation. In general, excluding some cases discussed in the literature [2, 6, 14, 19], for large times t one has $W_t \approx W_t'$ and relation(3.14) is also satisfied by W_t [10, 11, 17]. In Sect. (3.2.3.1) some numerical investigations of formula (3.8) are shown.

3.1.1.1 The Equilibrium Case: Kramers Equation

In order to verify the correctness of the equilibrium limit of formula (3.8) it is useful to test it on a generic equilibrium system. Let us consider N interacting particles $\{q_i, p_i\}$ coupled to a thermostat at temperature T with Hamiltonian[2] $H(\{q_i, p_i\}) = \sum_i \frac{p_i^2}{2m} + V(\{q_i\})$:

$$\begin{cases} \dot{q}_i &= \frac{\partial \mathcal{H}}{\partial p_i} \\ \dot{p}_i &= -\frac{\partial \mathcal{H}}{\partial q_i} - \gamma p_i + \eta_i(t), \end{cases} \tag{3.15}$$

where $\langle \eta_i(t)\eta_j(t')\rangle = 2\gamma T \delta_{i,j}\delta(t-t')$. Clearly, since the equilibrium distribution is given by the canonical one, namely $\rho \propto e^{-\beta\mathcal{H}}$ the border term is easily calculated to be $b_t = \beta\left(\mathcal{H}(0) - \mathcal{H}(t)\right)$.

Recalling that the momenta p_i are odd respect to the inversion of time, a straightforward application of Eqs. (3.3) and (3.4) gives

$$\begin{cases} D_i^{irr} &= -\gamma p_i \\ D_i^{rev} &= -\frac{\partial \mathcal{H}}{\partial q_i}. \end{cases} \tag{3.16}$$

Substituting it into (3.8) one obtains:

$$W_t \propto \int_0^t \left[\frac{\partial \mathcal{H}}{\partial p_i} \dot{p}_i + \frac{\partial \mathcal{H}}{\partial q_i} \dot{q}_i \right] dt. \tag{3.17}$$

Since the expression in the integrand is a total time derivative, it is straightforward to observe the border term $b_t = -W_t$ which clearly implies

$$W_t' \equiv 0, \tag{3.18}$$

[2] In order to lighten the notation, we consider the one-dimensional case. Clearly the results that follows are valid also for higher dimensions, once one substitute the products with scalar products.

for each trajectory separately, as expected from a system in equilibrium. Clearly, also S_{irr} is equal to zero.

Note that this result can be also extended to magnetic forces, which are not derived by a potential. For instance, let us consider the Lorentz force $\mathbf{F}_l$ acting on the particle i:

$$\mathbf{F}_l(\mathbf{x}) = \mathbf{B}(\mathbf{x}) \times \mathbf{p}_i, \tag{3.19}$$

where $\mathbf{B}$ is a non-uniform magnetic field. At the level of a classical description one must take into account that B is odd under time reversal and then $\mathbf{F}_l(\mathbf{x})$ is part of D^{rev}. The additional term in the entropy production due to the magnetic force is

$$W_M \propto \sum_i \int_0^t \mathbf{p}_i \cdot (\mathbf{B}(\mathbf{x}) \times \mathbf{p}_i) dt, \tag{3.20}$$

which vanishes for each particle i separately, thanks to the standard properties of the cross product.

In the overdamped limit, Eq. (3.15) becomes

$$\gamma \dot{q}_i = -\frac{\partial \mathcal{H}}{\partial q_i} + \eta(t), \tag{3.21}$$

and the role of the single terms changes: $D_{irr} = -\frac{1}{\gamma}\frac{\partial \mathcal{H}}{\partial q_i}$ and $D_{rev} = 0$. However, it is easy to observe that, also in this case, $W_t' = 0$ is easily obtained.

Let us consider now, from these examples, what are the main ingredients that make the equilibrium case easy to solve:

- The knowledge of the equilibrium distribution, given by the Boltzmann factor $e^{-\beta\mathcal{H}}$, allowing a correct computation of the border terms b_t.
- The separability of the Hamiltonian in a kinetic and potential part. In this way W_t is given by a total time derivative.
- The presence of an equilibrium thermostat, with the same energy scale for all the particles.

Moreover, W_t and b_t, at equilibrium, are indistinguishable, being both not increasing in time: in a sense, the detailed balance condition puts the dynamics on the same level of the statics.

It is easy to understand that no considerations of the same generality can be made for a generic out of equilibrium system. Moreover, the work in this case is harder because the steady state distribution is generally not known and from that it is in general not possible to derive the dynamical behavior of the system. However in quite all the cases, $W_t \sim \mathcal{O}(t)$ and $b_t \sim \mathcal{O}(1)$, the first one being the signal of the presence of a current flowing in the system, and the second one showing that a steady state is reached.

3.1.2 Irreversible Effects of Memory

One possible application of the description above is the presence of memory effects.

We present here the generalized Langevin equations with memory, considering the following equation of motion (we restrict ourselves to the one-dimensional problem, without loss of generality):

$$\begin{cases} \dot{x} = v \\ \dot{v} = F(x) - \int_{-\infty}^{t} \gamma(t-t')v(t')dt' + \eta(t), \end{cases} \tag{3.22}$$

with

$$\gamma(t) = 2\gamma_0\delta(t) + \sum_{i=1}^{M} \frac{\gamma_i}{\tau_i} e^{-\frac{t}{\tau_i}} \qquad \langle \eta(t) \rangle = 0 \tag{3.23}$$

$$\langle \eta(t)\eta(t') \rangle = 2T_0\gamma_0\delta(t-t') + \sum_{i=1}^{M} T_i \frac{\gamma_i}{\tau_i} e^{-\frac{|t-t'|}{\tau_i}}, \tag{3.24}$$

and where F is a generic drift term which can take the form of a sum of conservative and non-conservative forces, i.e., $F(x) \equiv -\frac{dU_0(x)}{dx} + F_{nc}(x)$. When $T_i = T_0$ for all i, the fluctuation dissipation relation of the second kind holds, as previously discussed in Sect. 2.1.1. The same analysis can be also carried out for the overdamped case [15].

This model has several applications: among others, it has been initially proposed in [4] for weakly driven glassy systems. More recently the noise in feedback cooled oscillator for gravitational wave detectors [1] has been characterized in a similar fashion [5]. In Chap. 4 we will argue that the dynamics of a tracer particle in moderately dense fluidized granular media, including its linear response properties, are consistent with this model. Note that the pairing of equal characteristic times for the exponentials in (3.23) and (3.24) is not so restrictive: indeed, case $T_i = 0$ or case $\gamma_i \to 0$, $T_i \to \infty$ with finite $\gamma_i T_i$, for some i, can be easily worked out and make no exception to the following analysis.

In order to use formula (3.8), it is necessary to map Eq. (3.22) into (3.1), where all noises are uncorrelated, identifying $N = M+2$ and $X_0 \equiv v$, $X_{N-1} \equiv x.X_i \equiv v_i$ ($i \in [1, M]$) are auxiliary variables necessary to take into account memory, for instance they can be defined as

$$v_i(t) = \sqrt{\frac{\gamma_i}{\tau_i}} \int_{-\infty}^{t} e^{-\frac{t-t'}{\tau_i}} \left(v(t') + \sqrt{\frac{T_i}{\gamma_i}} \xi_i(t') \right) dt'. \tag{3.25}$$

This sort of "Markovian embedding" is not a simple mathematical trick but it has a physical meaning. We will return to this point in Sect. 3.3.2.

With this choice for the auxiliary variables, it is easy to verify that the drifts in Eq. (3.1) are

$$D_0 = F(x) - \gamma_0 v - \sum_{i=1}^{M} \sqrt{\frac{\gamma_i}{\tau_i}} v_i \tag{3.26}$$

$$D_i = \sqrt{\frac{\gamma_i}{\tau_i}} v - \frac{1}{\tau_i} v_i \quad (i \in [1, M]) \tag{3.27}$$

$$D_{N-1} = v, \tag{3.28}$$

while the diagonal diffusion matrix reads

$$D_{00} = \gamma_0 T_0 \quad D_{ii} = \frac{T_i}{\tau_i}. \tag{3.29}$$

Summarizing, the system with memory is recast into a system of (linearly) coupled Langevin equations where all noises are uncorrelated. Auxiliary variables v_i ($i \in [1, M]$) are even under time-reversal, i.e., $\epsilon_i = 1$ for $i > 0$: this can be understood, for instance, requiring the validity of detail balance in the equilibrium case $T_i = T_0$ for all i.

Then we obtain

$$D_0^{rev} = F - \sum_{i=1}^{M} \sqrt{\frac{\gamma_i}{\tau_i}} v_i, \quad D_0^{ir} = -\gamma_0 v \tag{3.30}$$

$$D_i^{rev} = \sqrt{\frac{\gamma_i}{\tau_i}} v, \quad D_i^{ir} = -\frac{v_i}{\tau_i} \quad (i \in [1, M]) \tag{3.31}$$

$$D_{N-1}^{rev} = v, \qquad D_{N-1}^{ir} = 0. \tag{3.32}$$

When computing (Eq. 3.8), it is crucial to note that D_{ij} is not positive definite and cannot be inverted. Anyway, as noted by Machlup and Onsager [12], the last row and column of D_{ij} (those associated to variable x, which has not explicit noise dependence) can be dropped out for the purpose of computing path probabilities. With this observation, formula (3.8) can be used, leading to

$$W_t = -\sum_{i=0}^{M} \frac{\delta(v_i^2)}{2T_i} - \frac{\delta U_0}{T_0} +$$

$$+ \int_0^t \frac{1}{T_0} \left(F_{nc}[x(s)] + \sum_i F_i[v_i(s)] \right) v(s) ds \tag{3.33}$$

$$F_i = -\sqrt{\frac{\gamma_i}{\tau_i}} \left(1 - \frac{T_0}{T_i}\right) v_i(s). \tag{3.34}$$

As usual, exact differences appear, denoted as $\delta(g) \equiv g(t) - g(0)$. The non-trivial part of W_t is the time-integral on the right hand side of (3.33), which does not reduce to exact differences: it is equivalent to the work done by the usual non-conservative external force $F_{nc}(x)$ and by new forces F_i expressed in (3.34).

It can be verified that memory forces do not depend on the definition of auxiliary variables, as expected. The additional work done by the forces F_i is due to feedback of past history on the particle velocity and it is interesting to discover its effect on irreversibility. From formula (3.33) it is also evident that the force F_i vanishes if $T_i = T_0$. If $T_i = T_0$ for all i, then the fluctuation dissipation relation of the second kind holds, and memory does not contribute to W_t.

3.2 The Linear Model and Physical Interpretations

In this section we will apply the concept derived to the following "two variables" linear model, which has the advantage of being fully tractable. We consider two degrees of freedom X_1 and X_2

$$\begin{aligned}\dot{X}_1 &= -\alpha X_1 + \lambda X_2 + \sqrt{2D_1}\phi_1 \\ \dot{X}_2 &= -\gamma X_2 + \mu X_1 + \sqrt{2D_2}\phi_2,\end{aligned} \tag{3.35}$$

where $\alpha, \gamma, \lambda, \mu$ are constant coefficient and ϕ_1 and ϕ_2 are uncorrelated white noises, with zero mean and unitary variance.

The above stochastic equations can be thought as modelling the system portrayed in Fig. 3.1. The system includes two particles (for simplicity in one dimension), with positions x_1 and x_2 and momenta p_1 and p_2 whose Hamiltonian is given by

$$\mathcal{H}_{tot} = \frac{p_1^2}{2m_1} + \frac{p_2^2}{2m_2} + \frac{1}{2}k_1 x_1^2 + \frac{1}{2}k_2 x_2^2 + \frac{1}{2}k(x_1 - x_2)^2. \tag{3.36}$$

Each particle i is moving in a dilute fluid which exerts a viscous drag with coefficient γ_i, and which is coupled to a thermostat with temperature T_i; a tentative modelling for the dynamics of this system is the following:

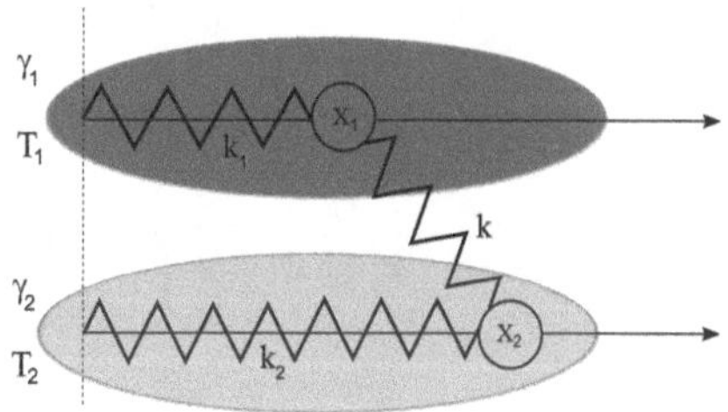

Fig. 3.1 graphical representation of the system described by (3.38)

$$\dot{p_1} = -\frac{\partial \mathcal{H}}{\partial x_1} - \gamma_1 \dot{x_1} + \sqrt{2\gamma_1 T_1}\phi_1$$
$$\dot{p_2} = -\frac{\partial \mathcal{H}}{\partial x_2} - \gamma_2 \dot{x_2} + \sqrt{2\gamma_2 T_2}\phi_2. \tag{3.37}$$

Now, by taking the overdamped limit we get:

$$\gamma_1 \dot{x_1} = -(k + k_1)x_1 + kx_2 + \sqrt{2\gamma_1 T_1}\phi_1$$
$$\gamma_2 \dot{x_2} = kx_1 - (k + k_2)x_2 + \sqrt{2\gamma_2 T_2}\phi_2, \tag{3.38}$$

which corresponds to model (3.35) by identifying $X_1 \to x_1$, $X_2 \to x_2$ and

$$\alpha \to \frac{k + k_1}{\gamma_1} \quad \lambda \to \frac{k}{\gamma_1} \quad \gamma \to \frac{k + k_2}{\gamma_2} \quad \mu \to \frac{k}{\gamma_2} \quad D_1 \to \frac{T_1}{\gamma_1} \quad D_2 \to \frac{T_2}{\gamma_2}. \tag{3.39}$$

In some cases it is helpful to consider an alternative interpretation of (Eq.3.35): this is realized, for instance, when considering a massive granular intruder in a gas of other granular particles driven by a stochastic external energy injection. Indeed the steady state dynamics of the intruder velocity $V \equiv X_1$ is fairly modelled by the following equation (as we will see in the chapter 4):

$$M\dot{V} = -\Gamma(V - U) + \sqrt{2\Gamma T_g}\phi_1$$
$$M'\dot{U} = -\Gamma' U - \Gamma V + \sqrt{2\Gamma' T_b}\phi_2, \tag{3.40}$$

where M is the intruder mass, Γ is the drag coefficient of the surrounding granular fluid, T_g is the granular temperature of the fluid, U is an effective local field interacting with the intruder, M' and Γ' are two parameters which characterize the effective mass and drag of the auxiliary field U, and finally T_b is the temperature of the external bath which keeps steady the system. This model is cast to model (3.38) by identifying $X_1 \to V$, $X_2 \to U$ and mapping the parameters in the following way:

$$\alpha \to \frac{\Gamma}{M} \quad \lambda \to \frac{\Gamma}{M} \quad \gamma \to \frac{\Gamma'}{M'} \quad \mu \to -\frac{\Gamma}{M'} \quad D_1 \to \frac{\Gamma T_g}{M^2} \quad D_2 \to \frac{\Gamma' T_b}{(M')^2}. \tag{3.41}$$

Note that in this interpretation the main variable X_1 is a velocity and therefore is odd under time-reversal (while it was even in the overdamped case); note also that μ here is negative. This two models possess the same mathematical structure, and can be treated in a similar manner. Also if they model different physical context, the underlying non-equilibrium mechanism is evidently the same: the presence of more than one energy source, which put the system out of equilibrium and make it different, for instance, from the case given by Eq. (3.15) or from the derivation presented in Sect. 2.1.1.

3.2.1 Steady State Properties

Model (3.35), in a more compact form, reads

$$\frac{d\mathbf{X}}{dt} = -A\mathbf{X} + \boldsymbol{\phi}, \tag{3.42}$$

where $\mathbf{X} \equiv (X_1, X_2)$ e $\boldsymbol{\phi} \equiv (\phi_1, \phi_2)$ are bidimensional vectors and A is a real 2×2 matrix, in general not symmetric. $\boldsymbol{\phi}(t)$ is a Gaussian process, with covariance matrix:

$$\langle \phi_i(t')\phi_j(t)\rangle = D_{ij}\delta(t - t'), \tag{3.43}$$

and

$$A = \begin{pmatrix} \alpha & -\lambda \\ -\mu & \gamma \end{pmatrix} \qquad D = \begin{pmatrix} 2D_1 & 0 \\ 0 & 2D_2 \end{pmatrix} \tag{3.44}$$

In order to reach a steady state, the real parts of A's eigenvalues must be positive. This condition is verified if $\alpha + \gamma > 0$ and $\alpha\gamma - \lambda\mu > 0$. Extension to a generic dimension d and non-diagonal matrices D (which however must remain symmetric) is straightforward.

The steady state is characterized by a bivariate Gaussian distribution:

$$\rho(\mathbf{X}) = N \exp\left(-\frac{1}{2}\mathbf{X}\sigma^{-1}\mathbf{X}\right), \tag{3.45}$$

where N is a normalization coefficient and the matrix of covariances σ satisfies

$$D = A\sigma + \sigma A^T. \tag{3.46}$$

Solving this equation gives

$$\sigma = \begin{pmatrix} \frac{D_2\lambda^2 - D_1\mu\lambda + D_1\gamma(\alpha+\gamma)}{(\alpha+\gamma)(\alpha\gamma-\lambda\mu)} & \frac{D_2\alpha\lambda + D_1\gamma\mu}{(\alpha+\gamma)(\alpha\gamma-\lambda\mu)} \\ \frac{D_2\alpha\lambda + D_1\gamma\mu}{(\alpha+\gamma)(\alpha\gamma-\lambda\mu)} & \frac{D_1\mu^2 - D_2\lambda\mu + D_2\alpha(\alpha+\gamma)}{(\alpha+\gamma)(\alpha\gamma-\lambda\mu)} \end{pmatrix}. \tag{3.47}$$

In order to capture the physical meaning of (3.47), it is useful to analyze the specific case described by the mapping (3.41):

$$\sigma = \begin{pmatrix} \frac{T_b}{M} + \Theta\Delta T & \Theta\Delta T \\ \Theta\Delta T & \frac{T_g}{M'} + \frac{\Gamma}{\Gamma'}\Theta\Delta T \end{pmatrix}, \tag{3.48}$$

where we have introduced $\Theta = \frac{\Gamma\Gamma'}{(\Gamma+\Gamma')(M'\Gamma+M\Gamma')}$ and $\Delta T = T_b - T_g$. From this reparametrization emerges that, when $T_b = T_g$ the two variables are uncorrelated. Conceptually similar considerations are valid also for the overdamped case (3.39) [15, 21].

We shall see in the next section how this emerging correlation between different degrees of freedom is strictly connected with both the fluctuation-response "violations" and the entropy production.

3.2.2 The Response Analysis

Thanks to linearity of Eq. (3.35), the response properties of the system can be easily calculated

$$R(t) = e^{-At}, \tag{3.49}$$

where $\mathbf{R}(t) \equiv \overline{\frac{\partial x_i(t)}{\partial x_j(0)}}$.

Moreover, since the steady state distribution is completely known also for the non-equilibrium case, it is possible to apply the generalized response Eq. (2.18), obtaining:

$$R(t) = C(t)\sigma^{-1}, \tag{3.50}$$

where σ^{-1} is the inverse of (3.47) and $C_{ij}(t) \equiv \langle v_i(t)v_j(t)\rangle$.

Let us first start to discuss the underdamped case, focusing for simplicity on the variable V. Using the mapping (3.41) one has:

$$\frac{\overline{\delta V(t)}}{\delta V(0)} = \sigma^{-1}_{VV}\langle V(t)V(0)\rangle + \sigma^{-1}_{UV}\langle V(t)U(0)\rangle. \tag{3.51}$$

In the equilibrium case when $T_g = T_b$, one has that the Einstein relation is recovered. This is quite simple to observe, since $\sigma^{-1}_{UV} = 0$ and one has $R_{VV}(t) \equiv C_{VV}(t)$.

In the general case σ_{UV} differs from 0 and the mobility μ is not simply given by the integral of the autocorrelation of velocity. This apparent "violation" is restored only if all the couplings between the different degrees of freedom are taken into account.

Similar considerations can be proposed also for the overdamped case. The only difference is that, since the observables are positions, one must compare the response matrix with the derivative of the autocorrelation. By deriving respect to time Eq. (3.49), and substituting into (3.50) gives

$$R(t) = \dot{C}(t)(A\sigma)^{-1}, \tag{3.52}$$

where, in this case

$$A\sigma = \begin{pmatrix} \frac{T_1}{\gamma_1} & \Sigma\Delta T \\ -\Sigma\Delta T & \frac{T_2}{\gamma_2} \end{pmatrix} \tag{3.53}$$

with $\Sigma = \frac{k}{(k+k_2)\gamma_2+(k+k_1)\gamma_1}$ and, as usual $\Delta T = (T_1 - T_2)$. The similarity with Eq. (3.48) is now explicit: when $T_1 = T_2$, also in this case, $(A\sigma)_{12}$ vanishes and one recovers the known fluctuation dissipation relation $\mathbf{R}(t) \propto \dot{\mathbf{C}}(t)$.

3.2.3 Entropy Production

The Entropy production of the system can be calculated after having taken a decision on the parity of the variable upon the time reversal transformation. As already pointed out in Sect. 3.1.1, in the exact expression of the entropy production are present also border terms, which are not extensive in time. We do not include this terms in the calculations, since we are interested in the asymptotic expression. We start from the case (3.38) where both the variables, being positions, are even under time reversal. Using (3.8), the entropy production is calculated to be

$$W_t = \frac{1}{2D_1}\int_0^t dt' \left(\lambda X_2 \dot{X}_1 - \alpha X_1 \dot{X}_1\right) + \frac{1}{2D_2}\int_0^t dt' \left(\lambda\mu X_1 \dot{X}_2 - \gamma X_2 \dot{X}_2\right). \tag{3.54}$$

Note that the terms $\int dt X_1 \dot{X}_1$ and $\int dt X_2 \dot{X}_2$ are not extensive in time. Therefore, for large times, equation (3.54) can be recast into

$$W_t \simeq \left[\frac{\lambda}{D_1} - \frac{\mu}{D_2}\right]\int_0^t X_2 \dot{X}_1 dt'. \tag{3.55}$$

It is then possible to calculate the mean value of the entropy production rate

$$\begin{aligned}\lim_{t\to+\infty}\frac{1}{t}\langle W_t\rangle &\simeq \left[\frac{\lambda}{D_1} - \frac{\mu}{D_2}\right]\frac{1}{t}\int_0^t X_2 \dot{X}_1 dt' \\ &= \left[\frac{\lambda}{D_1} - \frac{\mu}{D_2}\right]\langle X_2 \dot{X}_1\rangle.\end{aligned} \tag{3.56}$$

Equation (3.56) can be closed by substituting the equation of motion (3.35) and the values of the static correlations (3.47), obtaining

$$\lim_{t\to+\infty}\frac{1}{t}\langle W_t\rangle = \frac{(D_2\lambda - D_1\mu)^2}{D_1 D_2(\alpha+\gamma)}. \tag{3.57}$$

In the case (3.39), the formula gives:

$$\lim_{t\to+\infty}\frac{1}{t}\langle W_t\rangle = \frac{(k)^2}{(k_1+k)\gamma_2 + (k+k_2)\gamma_1}\frac{\Delta T^2}{T_2 T_1}. \tag{3.58}$$

A quite identical procedure can be performed also for the underdamped case, taking care of the parity transformations $V \to -V$, $U \to U$ under time reversal operation. In this case one obtains for the general model

$$W_t \simeq \left[\frac{\alpha\lambda}{D_1} + \frac{\mu\gamma}{D_2}\right] \int_0^t X_1 X_2 dt'. \tag{3.59}$$

Making the substitution (3.41):

$$W_t \simeq \Gamma\left(\frac{1}{T_g} - \frac{1}{T_b}\right) \int_0^t V(t')U(t')dt'. \tag{3.60}$$

and its mean entropy production rate is

$$\frac{1}{t}\langle W_t \rangle = \frac{\Gamma\Theta(\Delta T)^2}{T_g T_b}. \tag{3.61}$$

A direct comparison between formulas (3.61) and (3.58) shows how the mean rate is always positive, as expected. Moreover it is zero at equilibrium and in other more trivial cases, namely when the dynamical coupling terms (Γ or k) go to zero. It can approach to zero also in the limit of time scale separation, but we will return on this point in sect. 3.3.1.

3.2.3.1 Numerical Verifications

We conclude this section, evaluating the so-called finite time (or transient) contribution b_t which must be added in order to verify the fluctuation relation at short times. Since the steady state distribution is given by (3.45), from (3.13) one obtains:

$$b_t = \sigma_{11}^{-1}\frac{\delta X_1^2}{2} + \sigma_{22}^{-1}\frac{\delta X_2^2}{2} - \sigma_{01}^{-1}[X_1(0)X_2(0) + X_1(t)X_2(t)]. \tag{3.62}$$

Let us note again that, when $T_1 = T_2$, the term W_t exactly cancels b_t: in particular, it appears that $W_t' = 0$, i.e. detailed balance is satisfied.

In Fig. 3.2 we show the probability density function $p(W_t)$, in the steady state, obtained numerically by integrating Eq. (3.35), for different choices of times. In the same figure we also show the validity of Eq. (3.59) for W_t at large times and W_t' at any time (see the difference between empty and gray circles), as well as the asymptotic convergence to the large deviation rate function.

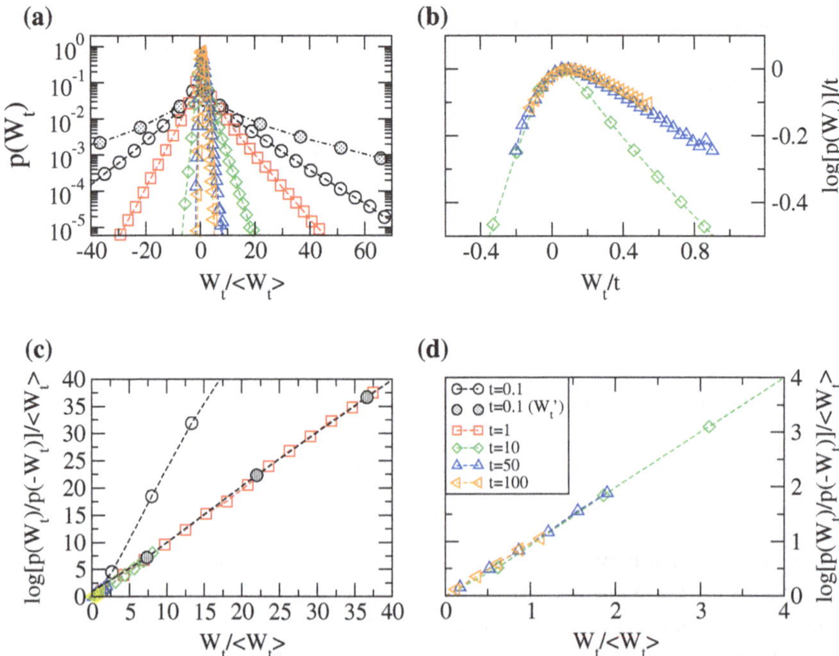

Fig. 3.2 **a** probability distribution of W_t for the inertial Langevin Eq. (3.35) i.e. $\alpha = 10$, $\lambda = \gamma = 1/\sqrt{2}$, $\mu = 0.1$, $D_1 = 6$, $D_2 = 0.3$, for different times t of integration. **b** at large times, $\log p(W_t/t)/t$ converges to a time-independent function: the large deviation rate. **c-d** check of the fluctuation relation (3.14) which is verified for all data aligned along the bisector. At small time (empty circles), where the fluctuation relation does not hold, Eq. (3.14) is verified for the probability distribution of W_t' (gray circles).

3.3 Entropy Production and Information

In order to predict the response of an equilibrium system it is sufficient to know only its autocorrelation, as stated from the fluctuation dissipation theorem. In a broad sense, autocorrelation and response have the same information content. On the contrary we have shown that the cross correlations between different degrees of freedom play a crucial role in the non-equilibrium cases. It appears natural to ask what is possible to conclude if one has access only to the main variable (x_1 or V), and, in a sense, forget the presence of the other "auxiliary" variables (x_2 or U).

From a mathematical point of view, it is equivalent to pass from a Markovian, namely from coupled equations with white noise, no memory effects and auxiliary variables, to a one-variable non-Markovian description, with memory and colored noise, doing the inverse respect to what done in Sect. 3.1.2.

In order to fix ideas, let us consider again the linear model (3.35). By integrating formally the second equation one has

$$X_2(t) = \int_{-\infty}^{t} dse^{-\gamma|t-s|}[\mu X_1(s) + \sqrt{2D_2}\phi_2(s)]. \tag{3.63}$$

Putting (3.63) into the equation for X_1 one obtains:

$$\dot{X}_1 = -\alpha X_1 + \lambda\mu \int_{-\infty}^{t} dse^{-\gamma|t-s|} X_1(s) + \eta(t) \tag{3.64}$$

with

$$\langle \eta(t)\eta(s) \rangle = 2D_1\delta(t-s) + \frac{D_2\lambda^2}{\gamma} e^{-\gamma|t-s|}. \tag{3.65}$$

It is worth noting that, with this mapping, the detailed balance condition, given in the Markovian description by the condition $D_1\mu = D_2\lambda$, is "translated" into

$$\langle \phi(t)\phi(s) \rangle \propto \Gamma(t-s), \tag{3.66}$$

which is the fluctuation dissipation relation of the second kind, for generalized langevin equations, as presented in Sect. 2.1.1.

This mapping appears to be an harmless mathematical trick, and one is tempted to consider both the processes at the same level. Actually it is related to a lost of information, detected by same observables, like entropy production, as we shall see in the next sections.

3.3.1 Auxiliary Variables Versus Effective Temperatures

In the last decades, in various systems exhibiting ergodicity breaking, the concept of effective temperature has been introduced, given by:

$$T_{eff}^{(AB)}(t, t_w) \equiv \frac{R_{AB}(t, t_w)}{\dot{C}_{AB}(t, t_w)}. \tag{3.67}$$

where A and B are two different observables of the system. Equation (3.67) represents an attempt to generalize the temperature in system out of equilibrium, where ergodicity is broken. The validity of a thermodynamic interpretation of this quantity is clear in some limits, namely well separated time-scales (see discussion in Sect. 2.2.1.4).

The purpose of this section is to answer to the following questions: What can we say about the system if we observe only the main variable? Is the effective temperature useful in this case?

At a first sight, Eq. (3.67) (considering, for instance, $A = B \equiv V$) appears in sharp contrast with the "cross-correlation" description given in Sect. 3.2.2, mainly because only the perturbed variable is involved. This difference is emphasized in

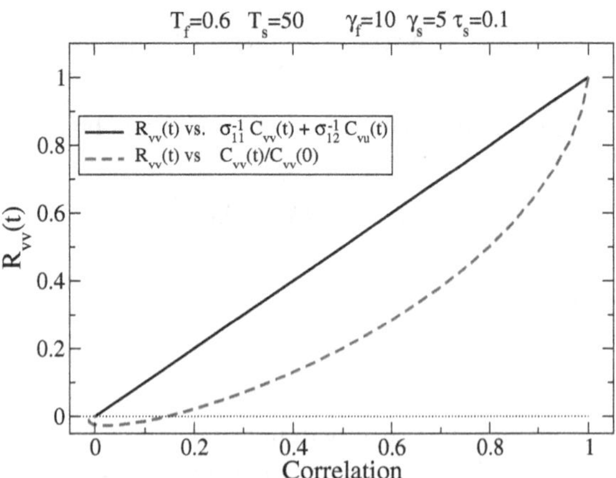

Fig. 3.3 Response of variable V in function of the unperturbed autocorrelation (red dashed line), when $T_g \neq T_b$. Linearity is restored if one use the generalized response function. The definition of these parameters can be found in [21]

Fig. 3.3, where the Einstein relation is violated: response $R_{VV}(t)$, when plotted against $C_{VV}(t)$, shows a non-linear relation. Anyway, a simple linear plot is restored when the response is plotted against the linear combination of correlations indicated by formula (3.51). In this case it is evident that the "violation" cannot be interpreted by means of any effective temperature, namely from this plot is not possible to extract the two temperatures underlying the dynamics. The violation is a consequence of having "missed" the coupling between variables V and U, which gives an additive contribution to the response of V.

However, in some cases also a partial view of the correlation response plot it is meaningful. In particular, this is again the case of time scales separation. For instance, let us take $M' \gg 1$ in (3.41). In this limit, the relaxing time τ_U of the variable U diverges. Roughly speaking, for time intervals lower than τ_U, one can consider $U(t)$ equal to the initial value U_0. Within this approximation the equation of the particle becomes:

$$M\dot{V} = -\Gamma(V - U_0) + \sqrt{2\Gamma T_g}\phi_1, \tag{3.68}$$

which is clearly the equation of a particle moving in a time independent force field. The response function is easily calculated:

$$R_{VV} = \frac{M}{T_g}\langle V(t)[V(0) - U_0]\rangle. \tag{3.69}$$

An interpretation of (3.69) is evident: it represent a sort of Einstein relation, between mobility and diffusion, calculated in the "Lagrangian frame" of the particle. Thanks to the time scale separation between the two variables, one can see that a linear relation is valid, apart from a slight modification. Note that this restoring of the fluctuation dissipation Relation in a Lagrangian frame has attracted some authors, and has inspired several works, both theoretical [3, 18] and experimental [7].

Also the model (3.38) has an interesting and non-trivial interplay of time-scales. For simplicity let us consider the case $k_2 = 0$. A typical time for variable x, corresponding to its relaxation time when decoupled by y (i.e. $k_2 = 0$), is $\tau_1 = \frac{\gamma_1}{k_1}$. Analogously it is possible to define a characteristic time for y: $\tau_2 = \frac{\gamma_2}{k}$. An interesting limit can be verified when these two time scales are well separated, e.g. when

$$\begin{aligned} &\tau_1 \ll \tau_2 \\ &k \sim k_1 \sim \mathcal{O}(1), \end{aligned}$$

where the additional second condition guarantees that the interactions have the same order of magnitude, so that the limit is non-trivial and remains of pure non-equilibrium. In this case it can be shown that the two timescales τ_1 and τ_2 correspond to those obtained by inverting the two eigenvalues of the matrix A. Most importantly, only in this limit the analysis of integrated response versus correlation produces a two slope curve, where T_1 and T_2 are recognized as inverse of the measured slopes. However this is a limit case, and more general conditions can be considered.

For a correct comparison with the effective temperature interpretation one must change slightly the definition of response used until this moment. Let us suppose to make a perturbation of the Hamiltonian (3.36) with a term $-h(t)x_1$. From equations of motion (3.37) one has

$$\overline{\frac{\delta x(t)}{\delta h(0)}} = \frac{1}{\gamma_1}\overline{\frac{\delta x(t)}{\delta x(0)}}. \tag{3.70}$$

The response of the system (3.52) can be recast into:

$$\overline{\frac{\delta x(t)}{\delta h(0)}} = \frac{(A\sigma)^{-1}_{11}}{\gamma_1}\frac{d}{dt}\langle x_1(t)x_1(0)\rangle + \frac{(A\sigma)^{-1}_{12}}{\gamma_1}\frac{d}{dt}\langle x_1(t)x_2(0)\rangle, \tag{3.71}$$

where the two contributions on the right side to the response depend on the time-scale of observation. In particular we consider the time-integrals of these two contributions, such that $\chi_{11}(t) = Q_{11}(t) + Q_{12}(t)$:

$$Q_{11}(t) = \frac{(A\sigma)^{-1}_{11}}{\gamma_1}[C_{11}(0) - C_{11}(t)] \tag{3.72}$$

$$Q_{12}(t) = \frac{(A\sigma)^{-1}_{21}}{\gamma_1}[C_{12}(0) - C_{12}(t)]. \tag{3.73}$$

Our choices of parameters are always with $T_1 \neq T_2$, as resumed in Table 3.1: a case (a) where the time-scales are mixed, and three cases (b), (c) and (d) where scales are well separated. In particular, in cases (c) and (d), the position of the intermediate plateau is shifted at one of the extremes of the parametric plot, i.e. only one range of time-scales is visible. Of course we do not intend to exhaust all the possibilities of this rich model, but to offer a few examples which are interesting for the following question: what is the meaning of the usual "incomplete" parametric plot χ_{11} versus C_{11}, which neglects the contribution of Q_{12}?

Table 3.1 Table of parameters for the 4 cases presented in Figs. 3.4 and 3.5. The effective time of the "fast" bath is defined as $\tau_1 = \gamma_1/k_1$, while the relaxation time of particle 2 is defined as $\tau_2 = \gamma_2/k$ (k_2 is always zero). The eigenvalues λ_1 and λ_2 of the dynamical matrix are also shown.

Case	T_2	T_1	γ_2	γ_1	τ_2	τ_1	k_1	k	$\frac{1}{\lambda_-}$	$\frac{1}{\lambda_+}$
a	5	0.2	20	40	30	20	2	2/3	47.3	12.7
b	2	0.6	200	1	200	1	1	1	400	0.5
c	2	0.6	100	100	2	1000	0.1	50	2000	1
d	10	2	1	50	10	50	1	0.1	51.2	9.76

The parametric plots, for the cases of Table 3.1, are shown in Fig. 3.4. In Fig. 3.5, we present the corresponding contributions $Q_{11}(t)$ and $Q_{12}(t)$ as functions of time. We briefly discuss the four cases:

(a) If the timescales are not separated, the general form of the parametric plot, see Fig. 3.4a, is a curve. In fact, as shown in Fig. 3.5a, the cross term $Q_{12}(t)$ is relevant at all the time-scales. The slopes at the extremes of the parametric plot, which can be hard to measure in an experiment, are $1/T_1$ and $s_\infty \neq 1/T_2$. Apart from that, the main information of the parametric plot is to point out the relevance of the coupling of x_1 with the "hidden" variable x_2.

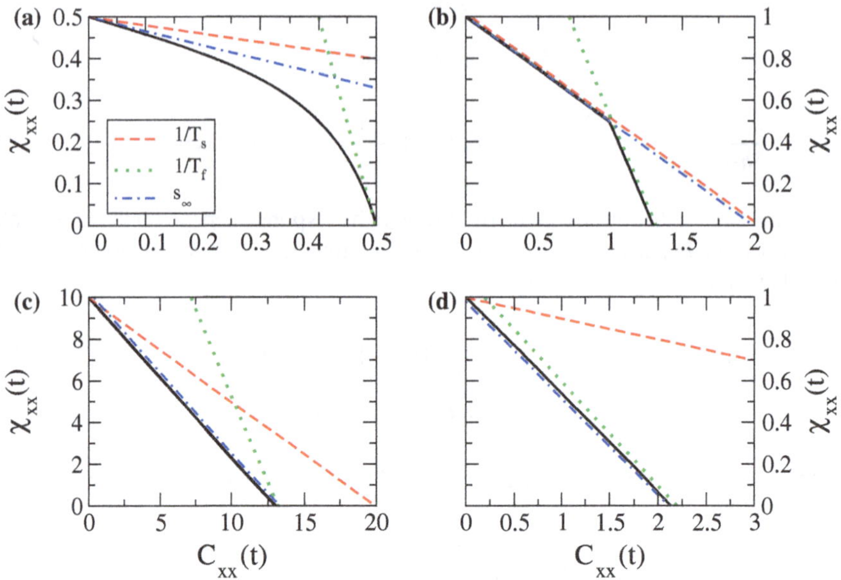

Fig. 3.4 Parametric plots of integrated response $\chi_{11}(t)$ versus self-correlation $C_{11}(t)$ for the model in Eq. (3.38) with parameters given in Table 3.1. Lines with slopes equal to $1/T_2$, $1/T_1$ and s_∞ are also shown for reference

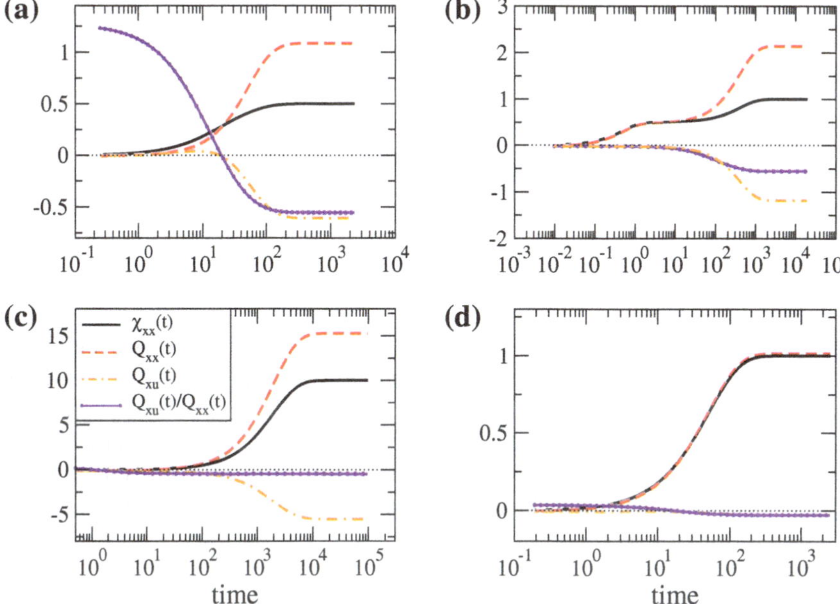

Fig. 3.5 Integrated response $\chi_{11}(t)$ as a function of time, for the model in Eq. (3.38) with parameters given in Table 3.1. The curves $Q_{11}(t)$ and $Q_{12}(t)$, representing the two contributions to the response, i.e. $\chi_{11}(t) = Q_{11}(t) + Q_{12}(t)$, are also shown. The violet curve with small circles represents the ratio $Q_{12}(t)/Q_{11}(t)$

(b) In the "glassy" limit $\tau_2 \gg \tau_1$, with the constraint $y_0 = \frac{T_1}{T_2}\frac{k_1}{k} \sim 1/2$, the well known broken line is found [22], see Fig. 3.4b, as discussed at the end of the previous section. Fig. 3.5b shows that $Q_{12}(t)$ is negligible during the first transient, up to the first plateau of $\chi(t)$, while it becomes relevant during the second rise of $\chi(t)$ toward the final plateau.
(c) If $\tau_1 \gg \tau_2$, the parametric plot, Fig. 3.4c, suggests an equilibrium-like behavior (similar to what one expects for $T_1 = T_2$) with an effective temperature $1/s_\infty$ which is different from both T_1 and T_2. Indeed, this case is quite interesting: the term Q_{12} is of the same order of Q_{11} during all relevant time-scales, but Q_{12}/Q_{11} appears to be almost constant. This leads to observe a plot with a non-trivial slope. The close similarity between Q_{11} and Q_{12} is due to the high value of the coupling constant k.
(d) In the last case, always with $\tau_1 \gg \tau_2$, the contribution of $Q_{12}(t)$ is negligible at all relevant time-scales (see Fig. 3.5), giving place to a straight parametric plot, shown in Fig. 3.4, with slope $1/T_1$. The low value of the coupling constant k is in agreement with this observation.

The lesson learnt from this brief study is that the shape of the parametric plot depends upon the timescales and the relative coupling. This is consistent with the fact that the correct formula for the response is always: $\frac{\overline{\delta x(t)}}{\delta h(0)} = \dot{Q}_{11} + \dot{Q}_{12}$. However, the definition of an effective temperature through the relation $T_{eff}(t)\frac{\overline{\delta x(t)}}{\delta h(0)} = \dot{Q}_{11}(t)$ in general (see case a), does not seem really useful. In particular limits, the behavior of the additional term Q_{12} is such that $R \propto \dot{Q}_{11}$ in a range of time-scales, and therefore the measure of T_{eff} becomes meaningful.

3.3.2 From the Non-Markovian to the Markovian Model

The previous section shows that if one takes the point of view of one variable a different interpretation of these fluctuation dissipation "violations" can be given, respect to the two variable case. This interpretations are not in contrast each other, namely the condition $T_1 = T_2$ is always the "equilibrium fingerprint" which satisfies the fluctuation dissipation theorem. The scenario is different if one compare the entropy production in the Non-Markovian system to what found in sect. 3.2.3.

Consider the following simple one-dimensional Langevin equation

$$m\ddot{x} = -\gamma\dot{x} - h_x[x(t)] - \int_{-\infty}^{t} dt'\, g(t-t')\, x(t') + \eta, \tag{3.74}$$

where $\eta(t)$ is Gaussian noise of zero mean and correlation and $h_x[x(t)]$ is a generic force.

$$\langle \eta(t)\, \eta(t') \rangle = \nu(t-t') \tag{3.75}$$

with $\nu(t) = \nu(-t)$. In this model one can calculate the path probability and its reversed. The mean-value of Lebowitz-Sphon functional, ignoring all the contributes non-extensive in time, reads

$$\left\langle \log \frac{\mathcal{P}\{x\}}{\mathcal{P}\{\mathcal{I}x\}} \right\rangle = -\int_{-\infty}^{+\infty} \frac{d\omega}{\pi}\, \omega \left\langle \mathrm{Im}\big[x(\omega) h_x(-\omega)\big] \right\rangle [\gamma + \psi(\omega)]\, \nu(\omega)^{-1}, \tag{3.76}$$

where the average is performed on the space of trajectories and $\mathcal{I}$ is time inversion operator. The functions appearing on the right side are the Fourier transforms (See Appendix A.2 for the details of the calculation).

From Eq. (3.76) it is easy to see that for the linear case, namely for $h_x(x) \propto x(t)$ one has:

$$\left\langle \mathrm{Im}\big[x(\omega) h_x(-\omega)\big] \right\rangle = 0. \tag{3.77}$$

Remarkably, it predicts a vanishing entropy production also in the case of the linear model for $T_1 \neq T_2$, in sharp contrast with what found for instance in (3.55) and (3.60).

From this result emerges that the two approaches represent the same physical situation but with different levels of details: moreover the choice of the level of the description does not affect almost any of the observables, for instance correlations and responses of the main variable are unaffected. In order to focus on the reason of this difference, let us consider the model with exponential memory (3.64) that we rewrite here in a lightened notation, for clarity:

$$\dot{x} = -h(x) + \lambda\mu \int_{t_0}^{t} e^{-\gamma(t-s)} x(s) + \eta(t) \equiv f_x + \eta(t). \tag{3.78}$$

The path probability of this process, starting form the position x_0 at time t_0, can be expressed in the following form (see Appendix A.3 for details)

$$P[x|x_0] = \int \mathcal{D}\sigma\, \delta[\dot{x} + f_x - \lambda y_0 g(t-t_0) - \lambda\sqrt{2D_y} \int_{t_0}^{t} ds g(t-s)\phi_y(s) - \sqrt{2D_x}\phi_x(t)] \tag{3.79}$$

where we have used a simplified notation $\{\mathbf{X}(s)\}_0^t \equiv x$ and we have defined $\mathcal{D}\sigma = dy_0 P_0(y_0)\mathcal{D}\mu[\phi_x]\mathcal{D}\mu[\phi_y]$ where the μ's are the Gaussian measures of the noises and P_0 is a Gaussian distribution with zero mean and variance $\frac{D_y}{\gamma}$.

After introducing an auxiliary process $\{\mathbf{Y}(s)\}_0^t \equiv y$, Eq. (3.79) can be recast into:

$$\begin{aligned} P[x|x_0] = \int \mathcal{D}\sigma \mathcal{D}y \delta[\dot{x} + h_x - \lambda y - \sqrt{2D_x}\phi_x(t)] \\ \times\ \delta[y - y_0 g(t-t_0) - \int_{t_0}^{t} ds g(t-s)[\mu x(s) + \sqrt{2D_y}\phi_y(s)]. \end{aligned} \tag{3.80}$$

After integrating over the noises, one obtains the following expression for the probability

$$P[x|x_0] = \int dy_0 P_0(y_0) \int_{y(0)=y_0} \mathcal{D}y e^{S(x,y)}, \tag{3.81}$$

where

$$S(x, y) = -\frac{1}{2D_x} \int_{t_0}^{t_1} dt[\dot{x} + h_x - \lambda y]^2 - \frac{1}{2D_y} \int_{t_0}^{t_1} dt[\dot{y} + \gamma y - \mu x]^2. \tag{3.82}$$

It is straightforward to recognize that Eq. (3.82) is the action of the corresponding two variable stochastic process:

$$\begin{cases} \dot{x} = -h_x + \lambda y + \sqrt{2D_x}\phi_x \\ \dot{y} = -\gamma y + \mu x + \sqrt{2D_y}\phi_y \end{cases} \tag{3.83}$$

for the particular choice of the initial condition y_0, following the Gaussian distribution P_0. This result shows how the probability distribution of the model (3.78) is essentially given by a marginalization of the corresponding Markovian one. From such an identification it is straightforward to explain the results shown in the previous sections.

3.3.3 Consequences of Projections

If we denote with $\langle \dots \rangle_x$ the average over the paths in the model (3.78) and with $\langle \dots \rangle_{x,y}$ the average on the equivalent model on the auxiliary variable, one has, [3] for an observable which depends only on x

$$\langle \mathcal{O} \rangle_x = \int \mathcal{D}x \mathcal{P}[x] \mathcal{O}(x) = \int \mathcal{D}x \mathcal{D}y e^{S(x,y)} \mathcal{O}(x) = \langle \mathcal{O} \rangle_{x,y}. \tag{3.84}$$

The relation (3.84) shows how, each observable of the variable x has the same values when computed in the two models.

On the contrary

$$\int \mathcal{D}x \mathcal{D}y \mathcal{P}[x,y] \log \left[\frac{\int \mathcal{D}y \mathcal{P}[x,y]}{\int \mathcal{D}y \mathcal{P}_1[x,y]} \right] \neq \int \mathcal{D}x \mathcal{D}y \mathcal{P}[x,y] \log \left[\frac{\mathcal{P}[x,y]}{\mathcal{P}_1[x,y]} \right]. \tag{3.85}$$

where we have defined with $P_1(x, y)$ the probability of the inverted trajectory.

And, as a consequence, $\langle W \rangle_x \neq \langle W \rangle_{x,y}$. This fact explains the difference observed. Moreover it is simple to observe that

$$\langle W \rangle_{x,y} - \langle W \rangle_x = \int \mathcal{D}x \mathcal{D}y \mathcal{P}[x,y] \log \frac{\mathcal{P}[x,y]}{\mathcal{P}_1[y|x]\mathcal{P}[x]} \geq 0, \tag{3.86}$$

where the last inequality is a straightforward application of the properties of Kullback-Leibler relative entropy, which is always non-negative [9]. Then, this projection mechanism, in general, has the effect of reducing entropy production. The equality is satisfied if

$$\mathcal{P}_1[y|x] = \mathcal{P}[y|x]. \tag{3.87}$$

The physical meaning of (3.87) is clear: it represents a sort of "reduced" detailed balance condition, it must be valid for the variables one wants to remove from the description.

If one removes variables which are in equilibrium with respect to the others degrees of freedom, the procedure will not affect the entropy production. It is simple to note that this condition is not valid for the model (3.38), once one decide to project away the variable y.

Under this point of view it is possible also to have an idea of why the projection mechanism is not dangerous when the time scales are well separated. Let us consider,

---

[3] we omit to write down the border terms contributions for simplicity.

for instance, the system in Fig. 3.1. In the limit of $\tau_2 \ll \tau_1$, the particle 1 can be seen as blocked. Therefore the particle 2 is in equilibrium respect to the system "thermostat + blocked particle x_1" and Eq. (3.87) is valid for every values of x.

3.3.3.1 An Example

To make clear the preceding discussion, let us consider a particular example where two different "channels" for entropy production can be put in evidence. The example consists in a particle subject to a non-equilibrium bath and to an external driving force F. The velocity follows the equation:

$$\dot{v} = -\int_{-\infty}^{t} \Gamma(t-t')v(t')dt' + F + \eta(t), \tag{3.88}$$

where

$$\Gamma(t-t') = 2\gamma_f \delta(t-t') + \frac{\gamma_s}{\tau_s} e^{-\frac{(t-t')}{\tau_s}} \tag{3.89}$$

$$\langle \eta(t)\eta(t') \rangle = 2T_f \gamma_f \delta(t-t') + T_s \frac{\gamma_s}{\tau_s} e^{-\frac{|t-t'|}{\tau_s}}. \tag{3.90}$$

Clearly, due to the presence of the non conservative force F, the particle reaches a non-zero average velocity:

$$V_{lim} = \frac{F}{\int_0^{+\infty} \Gamma(t)dt} = \frac{F}{\gamma_s + \gamma_f}. \tag{3.91}$$

In this case, also in the non-Markovian description, an entropy production rate does exist. Following formula (A.42) of the Appendix, such rate reads:

$$\sigma^{diss}(t) = F \int_{-\infty}^{t} dt' K(t-t') \left[v(t) + v(t') \right], \tag{3.92}$$

where

$$K(t) = \frac{\delta(t)}{T_f} + \frac{\gamma_s}{2\Omega T_f \gamma_f \tau_s^2} \left(1 - \frac{T_s}{T_f}\right) e^{-\Omega|t|}$$

$$\Omega = \frac{1}{\tau_s} \sqrt{\frac{\gamma_f T_f + \gamma_s T_s}{\gamma_f T_f}}.$$

The average entropy production rate can be exactly calculated, yielding to the following result:

$$\overline{\sigma^{diss}} = \left(\frac{1}{T_f} + \frac{\gamma_s}{T_f \gamma_f \tau_s^2 \Omega^2}\left(1 - \frac{T_s}{T_f}\right)\right) V_{lim} F$$
$$= \left(\frac{1}{T_f} + \frac{\gamma_s T_s}{\gamma_s T_s + \gamma_f T_f}\left(\frac{1}{T_s} - \frac{1}{T_f}\right)\right) V_{lim} F. \tag{3.93}$$

Such result clearly shows that the entropy production vanishes if the external driving F is removed; on the other side if $F \neq 0$ a production exists even if $T_f = T_s$.

The same calculation for the mean entropy production rate can be carried out also for the corresponding Markovian system, obtaining:

$$\overline{\sigma_M^{diss}} = \left(\frac{F}{T_f}\langle v \rangle + \gamma_s \left(\frac{1}{T_s} - \frac{1}{T_f}\right)\langle vu \rangle\right). \tag{3.94}$$

Note that, because of the driving force, the variable has non-zero mean. Therefore, by using $\langle vu \rangle = \langle \delta v \delta u \rangle + \langle u \rangle \langle v \rangle$, one obtains the following result for the average entropy production rate:

$$\overline{\sigma_M^{diss}} = \frac{(T_f - T_s)^2 \gamma_f \gamma_s}{(\gamma_s + \gamma_f)(1 + \gamma_f \tau_s) T_f T_s} + \left(\frac{1}{T_f} + \frac{\gamma_s}{\gamma_s + \gamma_f}\left(\frac{1}{T_s} - \frac{1}{T_f}\right)\right) V_{lim} F. \tag{3.95}$$

The first term in the sum (3.95) is completely absent in the non-Markovian approach, (Eq. 3.93), and is different from zero even if $F = 0$. The second term is slightly different from (3.93), where a weighted average on the temperatures is present in the prefactor. They become identical when $\gamma_s \gg \gamma_f$.

3.4 The Linear Channel for Energy Exchange

The linear Eq. (3.35) constitute a simplified model of a more complex, and perhaps realistic, system with $N \gg 1$ degrees of freedom: such system Σ is made of two subsystems, say Σ_1 and Σ_2, made of, respectively, N_1 and N_2 degrees of freedom, with $N_1 + N_2 = N$. The N_i degrees of freedom of subsystem Σ_i are coupled to a thermostat at temperature T_i and are immersed in an external confining potential, assumed harmonic for simplicity. Furthermore, the N_i degrees of freedom of subsystem Σ_i interact among themselves by intermolecular potential which are, in general, anharmonic. In each subsystem Σ_i there is also a probe with position x_i and momentum p_i, with mass much larger than all the others in the same subsystem: such condition on the masses of the probes is sufficient to expect a linear Langevin-like dynamics for this degree of freedom, where the (non-linear) interaction with all other molecules is represented by an uncorrelated noise, while a linear velocity drag is due to collisional relaxation, and of course the external harmonic potential is still present, reproducing the situation of Fig. 3.1 and Eq. (3.37). Finally, these two "slow" degrees of freedom (with respect to the faster and lighter molecules) are

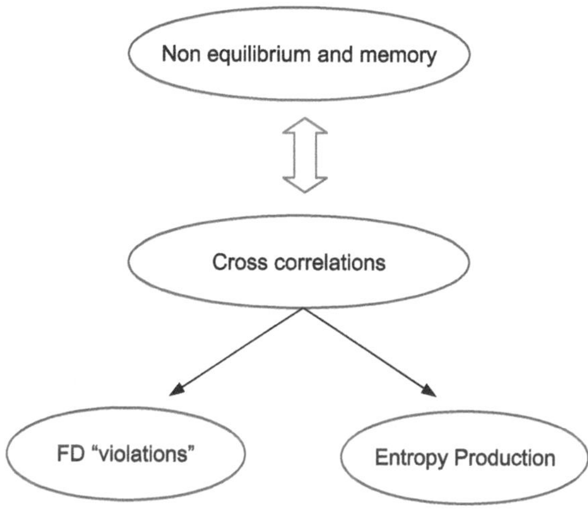

Fig. 3.6 Non equilibrium and memory effects produce non-trivial correlations among different degrees of freedom. Taking into account this aspect, it is possible to explain pure non equilibrium phenomena like fluctuation dissipation (FD) violations and entropy production

coupled one to the other by some potential $V(x_1 - x_2)$. This coupling is the only connection between systems Σ_1 and Σ_2.

In the absence of the coupling between the probes, the two systems remain separated and each one thermalizes to its own thermostat. When the coupling is present, the whole system will have the possibility to relax toward an overall equilibrium, but this is prevented by the presence of the two thermostats which are ideally infinite and never change their own temperature. The results is a non-equilibrium steady state where energy is continuously transferred on average from the hot to the cold reservoir.

However, in the linear case, the nature of the coupling may pose some ambiguities when the system is represented by the simplified 2-variables model. Indeed in the harmonic case the modes at different frequencies, i.e. $\tilde{X}_1(\omega)$, $\tilde{X}_2(\omega)$ will be decoupled. So, the only channel for heat to flow is the one connecting $\tilde{X}_1(\omega)$ to $\tilde{X}_2(\omega)$ with the same ω: the two components of the same mode are at different temperature and can exchange heat. In summary, each mode has its own channel, which is separated from the others. When the 2-variables model is reduced to the 1-variable model with memory, the information about this channel is completely lost because the two thermostats are reduced to only one. Each "cycle" at frequency ω which behaves as a loop with a given current, is flattened to a harmonic oscillator with zero net current. The only remaining entropy production belongs to the exchange between different modes, which cannot be present in the case of linear interaction and, as a consequence the entropy production vanishes. In this sense the single variable model does not faithfully reproduce the full entropy production of the whole system.

This mechanism of energy exchange is evidently given by the couplings between the different degrees of freedom (see Fig 3.6). The role of the coupling is crucial for two aspects:

- The response of the system to an impulsive perturbation is $R_{xx} = a\langle x(t)x(0)\rangle + b\langle x(t)y(0)\rangle$, where a and b are some constants. As expected, for the equilibrium limit ($T_1 \to T_2$), $b \to 0$ and the usual fluctuation response relation holds. On the contrary, when more than one thermostat is present, a coupling between different degrees of freedom emerges, "breaking" the usual form of the response relation.
- The entropy production rate can be calculated by using the Onsager-Machlup formalism. Also in this case, the rate is proportional to the cross correlations in which the factor depends on the two temperatures T_1 and T_2, and vanishes in the limit $T_1 \to T_2$.

These conclusions are not specific for the "two variables" model (3.35). As mentioned before also other variable can be inserted and the same description is still valid. Moreover, as shown in Sect. 3.1.2 also some non-linearity can be included in the description.

References

1. Bonaldi, M., et al.: Nonequilibrium steady-state fluctuations in actively cooled resonators. Phys. Rev. Lett **103**, 10601 (2009)
2. Bonetto, F., Gallavotti, G., Giuliani, A., Zamponi, F.: Chaotic hypothesis, fluctuation theorem, singularities. J. Stat. Phys. **123**, 39 (2006)
3. Chetrite, R., Gawędzki, K.: Eulerian and lagrangian pictures of non-equilibrium diffusions. J. Stat. Phys. **137**, 890 (2009)
4. Cugliandolo, L.F., Kurchan, J.: A scenario for the dynamics in the small entropy production limit. J. Phys. Soc. Jpn **69**, 247 (2000)
5. De Gregorio, P., Rondoni, L., Bonaldi, M., Conti, L.: Harmonic damped oscillators with feedback: a langevin study. J. Stat. Mech: Theory Exp **2009**, P10016 (2009)
6. Evans, D., Searles, D., Rondoni, L.: Application of the gallavotti-cohen fluctuation relation to thermostatted steady states near equilibrium. Phys. Rev. E **71**, 056120 (2005)
7. Gomez-Solano, J., Petrosyan, A., Ciliberto, S., Maes, C.: Fluctuations and response in a non-equilibrium micron-sized system. J. Stat. Mech: Theory Exp **2011**, P01008 (2011)
8. Hänggi, P. : Path integral solution for nonlinear generalized langevin equations. In: Proceedings of Path Integrals for meV to MeV: Tutzing'92, p.289 (1993)
9. Kullback, S., Leibler, R.: On information and sufficiency. Ann. Math. Stat **22**, 79 (1951)
10. Kumaran, V.: Temperature of a granular material "fluidized" by external vibrations. Phys. Rev. E **57**, 5660 (1998)
11. Lebowitz, J.L., Spohn, H.: A gallavotti-cohen-type symmetry in the large deviation functional for stochastic dynamics. J. Stat. Phys. **95**, 333 (1999)
12. Machlup, S., Onsager, L.: Fluctuations and irreversible process. II. Systems with kinetic energy. Phys. Rev. **91**, 1512 (1953)
13. Onsager, L., Machlup, S.: Fluctuations and irreversible processes. Phys. Rev. **91**, 1505 (1953)
14. Puglisi, A., Rondoni, L., Vulpiani, A. : Relevance of initial and final conditions for the fluctuation relation in markov processes. J. Stat. Mech. P08010 (2006)
15. Puglisi, A., Villamaina, D.: Irreversible effects of memory. Europhys. Lett. **88**, 30004 (2009)
16. Risken, H.: The Fokker-Planck Equation: Methods of Solution and Applications. Springer, Berlin (1989)
17. Seifert, U.: Entropy production along a stochastic trajectory and an integral fluctuation theorem. Phys. Rev. Lett. **95**, 040602 (2005)

18. Speck, T., Seifert, U.: Restoring a fluctuation-dissipation theorem in a nonequilibrium steady state. Europhys. Lett. **74**, 391 (2006)
19. van Zon, R., Cohen, E.G.D.: Extension of the fluctuation theorem. Phys. Rev. Lett. **91**, 110601 (2003)
20. van Zon, R., Cohen, E.G.D.: Stationary and transient work-fluctuation theorems for a dragged Brownian particle. Phys. Rev. E **67**, 046102 (2003)
21. Villamaina, D., Baldassarri, A., Puglisi, A., Vulpiani, A. Fluctuation dissipation relation: how to compare correlation functions and responses?. J. Stat. Mech. P07024 (2009)
22. Zamponi, F., Bonetto, F., Cugliandolo, L. F., and Kurchan, J. A fluctuation theorem for non-equilibrium relaxational systems driven by external forces. J. Stat. Mech. P09013 (2005)

Chapter 4
The Motion of a Tracer in a Granular Gas

Abstract This chapter is entirely devoted to a granular gas model. A Langevin equation for a massive tracer is obtained from the linear Boltzmann equation via a Kramers-Moyal expansion in a diluite limit. Such an expansion is not sufficient to observe non equilibrium effects and an equilibrium-like effective regime is obtained, without the presence of memory. In a denser regime, when the molecular chaos fails, the equation for a tracer is well represented by a Langevin equation with memory, and a local velocity field plays the role of an auxiliary variable coupled to the tracer. Finally, a strong assessment of the validity of the "local field" interpretation is given by the numerical verification of the fluctuation relation.

In the previous chapter we have shown the importance of memory effects and what is their role when combined with the breaking of detailed balance condition. Moreover we have underlined how a projection operation can produce relevant differences when dealing with entropy production calculations. In this chapter we will see an interesting application of this ideas.

In the first part we will analyze the transport properties of a massive intruder in a granular gas. The difference of masses between the intruder and the surrounding particles has the clear advantage to introduce a net time scale separation. This condition guarantees that the linear Boltzmann equation can be treated in the diffusive limit by applying the so-called Kramers-Moyal expansion.

As discussed below, the only knowledge of the tracer, in a molecular chaos approximation, it not sufficient to appreciate the non-equilibrium effects produced by the presence of inelasticity: this is an example how the projection has produced a loss of information, detected by entropy production. Starting from this result, the natural way to go on is to relax some condition: this is presented in Sect. 4.2, where the intruder has been studied in a denser regime, where the molecular chaos and, consequently, the linear Boltzmann equation are no more valid. Even in the absence of a general theory we show how a minimal correction to the dilute case is able to explain the main features of the system. The memory term due to the effect of recollision plays a central role for the prediction of the response properties. Remarkably, within

D. Villamaina, *Transport Properties in Non-Equilibrium and Anomalous Systems*, Springer Theses, DOI: 10.1007/978-3-319-01772-3_4,

this theory, a non equilibrium coupling emerges, completely absent in the case of a dense elastic fluid, which is the main source of entropy production of the system.

4.1 Diffusion of the Intruder in the Dilute Limit

4.1.1 Model

The model considered here is a slight variation of that one presented in Sect. 2.3.1: an "intruder" disc of mass $m_0 = M$ and radius R moves in a gas of N granular discs with mass $m_i = m$ $(i > 0)$ and radius r, in a two dimensional box of area $A = L^2$. We denote by $n = N/A$ the number density of the gas and by ϕ the occupied volume fraction, i.e. $\phi = \pi(Nr^2 + R^2)/A$ and we denote by $\boldsymbol{V}$ (or $\boldsymbol{v}_0$) and $\boldsymbol{v}$ (or $\boldsymbol{v}_i$ with $i > 0$) the velocity vector of the tracer and of the gas particles, respectively. Interactions among the particles are hard-core binary instantaneous inelastic collisions, such that particle i, after a collision with particle j, comes out with a velocity

$$\boldsymbol{v}_i' = \boldsymbol{v}_i - (1+\alpha)\frac{m_j}{m_i + m_j}[(\boldsymbol{v}_i - \boldsymbol{v}_j)\cdot\hat{\boldsymbol{\sigma}}]\hat{\boldsymbol{\sigma}}, \tag{4.1}$$

where $\hat{\boldsymbol{\sigma}}$ is the unit vector joining the particles' centers of mass and $\alpha \in [0, 1]$ is the restitution coefficient ($\alpha = 1$ is the elastic case). The mean free path of the intruder is proportional to $l_0 = 1/(n(r + R))$ and we denote by τ_c its mean collision time. Two kinetic temperatures can be introduced for the two species: the gas granular temperature $T_g = m\langle \boldsymbol{v}^2\rangle/2$ and the tracer temperature $T_{tr} = M\langle \boldsymbol{V}^2\rangle/2$.

In order to maintain a granular medium in a fluidized state, an external energy source is coupled to each particle in the form of a thermal bath [1–3] (hereafter, exploiting isotropy, we consider only one component of the velocities), so that the whole equation ruling the dynamics of a single particle is:

$$m_i\dot{v}_i(t) = -\gamma_b v_i(t) + f_i(t) + \xi_b(t). \tag{4.2}$$

Here $f_i(t)$ is the force taking into account the collisions of particle i with other particles, $\xi_b(t)$ is a white noise (different for all particles), with $\langle\xi_b(t)\rangle = 0$ and $\langle\xi_b(t)\xi_b(t')\rangle = 2T_b\gamma_b\delta(t-t')$. The effect of the external energy source balances the energy lost in the collisions so that a stationary state is attained with $m_i\langle v_i^2\rangle \le T_b$.

4.1.2 The Boltzmann Equation for the Tracer

Let us start by writing the coupled Boltzmann equations for the probability distributions of the gas particles, $P(\mathbf{v}, t)$ and the tracer one $P(\mathbf{V}, t)$, respectively [4]:

$$\frac{\partial P(\mathbf{V},t)}{\partial t} = \int d\mathbf{V}'[W_{tr}(\mathbf{V}|\mathbf{V}')P(\mathbf{V}',t) - W_{tr}(\mathbf{V}'|\mathbf{V})P(\mathbf{V},t)] + \mathcal{B}_{tr}P(\mathbf{V},t)$$

$$\frac{\partial p(\mathbf{v},t)}{\partial t} = \int d\mathbf{v}'[W_g(\mathbf{v}|\mathbf{v}')p(\mathbf{v}',t) - W_g(\mathbf{v}'|\mathbf{v})p(\mathbf{v},t)] + \mathcal{B}_g p(\mathbf{v},t) + J[\mathbf{v}|p,p], \tag{4.3}$$

where $\mathcal{B}_{tr}$ and $\mathcal{B}_g$ are two operators taking into account the interactions with the thermal bath. In these equations the effects of the collisions between the tracer and the gas particles are described by, respectively,

$$W_{tr}(\mathbf{V}|\mathbf{V}') = \chi \int d\mathbf{v}' \int d\hat{\sigma} p(\mathbf{v}',t)\Theta\left[-\left(\mathbf{V}'-\mathbf{v}'\right)\cdot\hat{\sigma}\right]\left(\mathbf{V}'-\mathbf{v}'\right)\cdot\hat{\sigma} \times \delta^{(d)}\left\{\mathbf{V}-\mathbf{V}'+\frac{\epsilon^2}{1+\epsilon^2}(1+\alpha)\left[\left(\mathbf{V}'-\mathbf{v}'\right)\cdot\hat{\sigma}\right]\hat{\sigma}\right\} \tag{4.4}$$

and

$$W_g(\mathbf{v}|\mathbf{v}') = \frac{\chi}{N} \int d\mathbf{V}' \int d\hat{\sigma} P(\mathbf{V}',t)\Theta\left[-\left(\mathbf{V}'-\mathbf{v}'\right)\cdot\hat{\sigma}\right]\left(\mathbf{V}'-\mathbf{v}'\right)\cdot\hat{\sigma} \times \delta^{(d)}\left\{\mathbf{v}-\mathbf{v}'+\frac{1}{1+\epsilon^2}(1+\alpha)\left[\left(\mathbf{v}'-\mathbf{V}'\right)\cdot\hat{\sigma}\right]\hat{\sigma}\right\}, \tag{4.5}$$

where $\Theta(x)$ is the Heaviside step function, $\delta^{(d)}(x)$ is the Dirac delta function in d dimensions, and $\chi = \frac{g_2(r+R)}{l_0}$, $g_2(r+R)$ being the pair correlation function for a gas particle and an intruder at contact; in the expressions (4.4) and (4.5) we have assumed that the probability $P_2\left(|\mathbf{x}-\mathbf{X}| = r+R, \mathbf{V}, \mathbf{v}, t\right)$ is given by the Enskog approximation (see Sect. 2.3.1.1)

$$P_2\left(|\mathbf{x}-\mathbf{X}| = r+R, \mathbf{V}, \mathbf{v}, t\right) = g_2(r+R)P(\mathbf{V},t)p(\mathbf{v},t), \tag{4.6}$$

which is a small correction to Molecular Chaos [5], taking into account density correlations near the intruder; the terms describing the action of the thermal bath read

$$\mathcal{B}_{tr} P(\mathbf{V},t) = \frac{\gamma_b}{M}\frac{\partial}{\partial \mathbf{V}}[\mathbf{V}P(\mathbf{V},t)] + \frac{\gamma_b T_b}{M}\Delta_V[P(\mathbf{V},t)] \tag{4.7}$$

$$\mathcal{B}_g p(\mathbf{v},t) = \frac{\gamma_b}{m}\frac{\partial}{\partial \mathbf{v}}[\mathbf{v}p(\mathbf{v},t)] + \frac{\gamma_b T_b}{m}\Delta_v[p(\mathbf{v},t)], \tag{4.8}$$

where Δ_v is the Laplacian operator with respect to the velocity. The last term on the right in the equation for $P(\mathbf{v},t)$ (4.3) is the particle-particle collisional term, $J[\mathbf{v}|p,p]$, which has a standard expression not reported here [6].

4.1.2.1 Decoupling the Gas from the Tracer

The goal of this section is to obtain, in some limit, a simple equation for the motion of the tracer in the "sea" of inelastic particles. The main problem which rise on, is that, in general, Eqs. (4.5) and (4.4) are coupled through with the transitions rate. In order to achieve this decoupling, one can choose the parameters of the system in order to have τ_c^g and τ_c^{tr}, respectively the gas and the tracer mean collision time, such that $\tau_c^g \ll \tau_c^{tr}$. This happens when the number N of the particles is large: in this case $P(\mathbf{V}, t)$ and $P(\mathbf{v}, t)$ change on well separated time scales. This time scale separation is confirmed by observing that $W_g(\mathbf{v}|\mathbf{v}')$ is $\mathcal{O}(N^{-1})$ and $W_{tr}(\mathbf{V}|\mathbf{V}')$ is $\mathcal{O}(1)$. The gas is weakly perturbed by the presence of the tracer and then one can assume that the probability $p(v)$ is independent on time. It is well known that the velocity distribution of a granular gas is not given by the Maxwellian and corrections to gaussianity in forms of Sonine polynomials [7] must be taken into account. Since we are in a dilute regime, we decide to neglect the corrections, supported by numerical evidences. The velocity distribution takes then the form:

$$p(\mathbf{v}) = \frac{1}{(2\pi T_g/m)^d} \exp\left[-\frac{m\mathbf{v}^2}{2T_g}\right]. \tag{4.9}$$

The main advantage of this assumption is evident by observing that the evolution Eq. (4.4), once (4.9) is inserted in, becomes Markovian and linear in $P(\mathbf{V}, t)$: a master equation for the evolution of the tracer is obtained.

4.1.3 The Kramers Moyal Expansion

With the assumptions discussed above, the linear master equation for the tracer is

$$\frac{\partial P(\mathbf{V}, t)}{\partial t} = L_{\text{gas}}[P(\mathbf{V}, t)] + L_{\text{bath}}[P(\mathbf{V}, t)], \tag{4.10}$$

where $L_{gas}[P(\mathbf{V}, t)]$ is a linear operator which can be expressed by means of the Kramers-Moyal expansion [8]

$$L_{\text{gas}}[P(\mathbf{V}, t)] = \sum_{n=1}^{\infty} \frac{(-1)^n \partial^n}{\partial V_{j_1} \dots \partial V_{j_n}} D^{(n)}_{j_1 \dots j_n}(\mathbf{V}) P(\mathbf{V}, t), \tag{4.11}$$

(the sum over repeated indices is meant) with

$$D^{(n)}_{j_1 \dots j_n}(\mathbf{V}) = \frac{1}{n!} \int d\mathbf{V}' (V'_{j_1} - V_{j_1}) \dots (V'_{j_n} - V_{j_n}) W_{tr}(\mathbf{V}'|\mathbf{V}), \tag{4.12}$$

and W_{tr} is given by relation (4.4). The second term in the master equation represents the interaction with thermal bath:

$$L_{\text{bath}}[P(\mathbf{V},t)] = \mathcal{B}_{tr} P(\mathbf{V},t). \tag{4.13}$$

In the limit of large mass M, we expect that the interaction between the granular gas and the tracer can be described by means of an effective Langevin equation. In this case, we keep only the first two terms of the expansion [8]

$$L_{\text{gas}}[P(\mathbf{V},t)] = -\frac{\partial}{\partial V_i}[D_i^{(1)}(\mathbf{V})P(\mathbf{V},t)] + \frac{\partial^2}{\partial V_i \partial V_j}[D_{ij}^{(2)}(\mathbf{V})P(\mathbf{V},t)]. \tag{4.14}$$

It is useful at this point to introduce the velocity-dependent collision rate and the total collision frequency

$$r(\mathbf{V}) = \int d\mathbf{V}' W_{tr}(\mathbf{V}'|\mathbf{V}), \tag{4.15}$$

$$\omega = \int d\mathbf{V}\ P(\mathbf{V}) r(\mathbf{V}). \tag{4.16}$$

The collision rate can be exactly calculated, giving

$$r(\mathbf{V}) = \chi\sqrt{\frac{\pi}{2}}\left(\frac{T_g}{m}\right)^{1/2} e^{-\epsilon^2 q^2/4} \times \left[(\epsilon^2 q^2 + 2) I_0\left(\frac{\epsilon^2 q^2}{4}\right) + \epsilon^2 q^2 I_1\left(\frac{\epsilon^2 q^2}{4}\right)\right], \tag{4.17}$$

where the rescaled variable $\mathbf{q} = \mathbf{V}/\sqrt{T_g/M}$, $\epsilon = m/M$ and $I_n(x)$ are the modified Bessel functions. To have an approximation of ω, on the other hand, one must do an assumption on $P(\mathbf{V})$. Let us take it to be a Gaussian with variance T_{tr}/M. The consistency of this choice will be verified in the following section. With this assumption, the collision rate turns out to be

$$\omega = \chi\sqrt{2\pi}\sqrt{T_g/m + T_{tr}/M} = \chi\sqrt{2\pi}\left(\frac{T_g}{m}\right)^{1/2}\sqrt{1 + \frac{T_{tr}}{T_g}\epsilon^2} = \omega_0 K(\epsilon), \tag{4.18}$$

where $\omega_0 = \chi\sqrt{2\pi}\left(\frac{T_g}{m}\right)^{1/2}$ and $K(\epsilon) = \sqrt{1 + \frac{T_{tr}}{T_g}\epsilon^2}$.

4.1.3.1 The Large Mass Limit of the Tracer

At this stage we can compute the terms $D_i^{(1)}$ and $D_{ij}^{(2)}$ appearing in L_{gas}. The result and the details of the computation of these coefficients as functions of ϵ are given in Appendix A.4. Here, in order to be consistent with the approximation in (4.14), from

Eq. (A.74) we report only terms up to ϵ^4

$$
\begin{aligned}
D_x^{(1)} &= -\chi\sqrt{2\pi}\frac{T_g}{m}q_x(1+\alpha)\epsilon^3 + \mathcal{O}(\epsilon^5) \\
&= -\chi\sqrt{2\pi}\left(\frac{T_g}{m}\right)^{1/2}(1+\alpha)\epsilon^2 V_x + \mathcal{O}(\epsilon^5) \\
&= -\omega_0(1+\alpha)\epsilon^2 V_x + \mathcal{O}(\epsilon^5) \qquad (4.19)\\
D_y^{(1)} &= -\omega_0(1+\alpha)\epsilon^2 V_y + \mathcal{O}(\epsilon^5) \qquad (4.20)\\
D_{xx}^{(2)} = D_{yy}^{(2)} &= \chi\sqrt{\pi/2}\left(\frac{T_g}{m}\right)^{3/2}(1+\alpha)^2\epsilon^4 + \mathcal{O}(\epsilon^5) \\
&= \frac{\omega_0}{2}\frac{T_g}{m}(1+\alpha)^2\epsilon^4 + \mathcal{O}(\epsilon^5) \qquad (4.21)\\
D_{xy}^{(2)} &= \mathcal{O}(\epsilon^6). \qquad (4.22)
\end{aligned}
$$

The linear dependence of $D_\beta^{(1)}$ upon V_β (for each component β), allows a granular viscosity

$$\eta_g = \omega_0(1+\alpha)\epsilon^2. \qquad (4.23)$$

In the elastic limit $\alpha \to 1$, one retrieves the classical results: $\eta_g \to 2\omega_0\epsilon^2$ and $D_{xx}^{(2)} = D_{yy}^{(2)} \to 2\omega_0\epsilon^2\frac{T_g}{M}$. In this limit the fluctuation dissipation relation of the second kind is satisfied [9, 10], namely the ratio between the noise amplitude and the drag coefficient γ_g, associated to the same source (collision with gas particles), is exactly T_g/M. When the collisions are inelastic, $\alpha < 1$, one sees two main effects: (1) the time scale associated to the drag $\tau_g = 1/\eta_g$ is modified by a factor $\frac{1+\alpha}{2}$, that is it is weakly influenced by inelasticity; (2) the Fluctuation-Dissipation relation of the second kind is violated by the same factor $\frac{1+\alpha}{2}$. This is only a partial conclusion, which has to be re-considered in the context of the full dynamics, including the external bath: this is discussed in the next section.

4.1.3.2 Langevin Equation for the Tracer

By taking the terms of the expansion up to the fourth order in ϵ, we are finally able to write the Langevin equation for the tracer

$$M\dot{\mathbf{V}} = -\Gamma_E\mathbf{V} + \mathcal{E}, \qquad (4.24)$$

where $\Gamma_E = \gamma_b + \gamma_g$ (the subscript E highlights that is derived under the Enskog approximation (4.6)) and $\mathcal{E} = \boldsymbol{\xi}_b + \boldsymbol{\xi}_g$, with

$$\gamma_g = M\eta_g = M\omega_0(1+\alpha)\epsilon^2 = \omega_0(1+\alpha)m \qquad (4.25)$$

$$\langle \mathcal{E}_i(t)\mathcal{E}_j(t')\rangle = 2\left[\gamma_b T_b + \gamma_g\left(\frac{1+\alpha}{2}T_g\right)\right]\delta_{ij}\delta(t-t'), \tag{4.26}$$

concluding that the stationary velocity distribution of the intruder is Gaussian with temperature

$$T_{tr} = \frac{\gamma_b T_b + \gamma_g\left(\frac{1+\alpha}{2}T_g\right)}{\gamma_b + \gamma_g}. \tag{4.27}$$

Equation (4.24) is linear in V, confirming the validity of the Gaussian *ansatz* used in computing ω_0. Note that the above expression for T_{tr} is consistent with the large mass expansion obtained in Eq. (4.22) only if it is dominated by T_g, for instance when $\gamma_g \gg \gamma_b$ (see discussion at the end of A.4). In the opposite limit, the tracer dynamics is dominated by the coupling with the external bath and the typical velocity of the tracer cannot be taken sufficiently small with respect to the typical velocity of gas particles, making the expansion unreliable.

For the self-diffusion coefficient it is immediately obtained

$$D_{tr} = \int_0^\infty dt\langle V_x(t)V_x(0)\rangle = \frac{T_{tr}}{\Gamma_E} = \frac{\gamma_b T_b + \gamma_g\left(\frac{1+\alpha}{2}T_g\right)}{(\gamma_b+\gamma_g)^2}, \tag{4.28}$$

which is largely verified in numerical simulations [11].

Other studies on different models of driven granular gases have found expressions very close to Eq. (4.25), which is not surprising considering the universality of the main ingredient for this quantity: the collision integral [12, 13].

4.1.4 The Motion of the Tracer as an Equilibrium-Like Process

Before concluding this section, let us summarize the approximations used so far to obtain the Langevin equation for the tracer:

1. *Molecular chaos approximation*: this is supposed to be true in dilute cases, where recollision phenomena, the main source of molecular chaos breaking, are not relevant. The Enskog correction implemented here does not changes this scenario.
2. *Large number of gas particles*: from a intuitive point of view, this condition allows to consider a "sea" of gas particles, which are weakly perturbed by the presence of the intruder. From a mathematical point of view, thanks to this limit, as evident in (4.5), $W_g(\mathbf{v}|\mathbf{v}')$ is $\mathcal{O}(1/N)$ and can be neglected.
3. *Gaussian approximation for the gas particles*: in principle, Eq. (4.3), also if one neglects $W_g(\mathbf{v}|\mathbf{v}')$, is solved by using the Sonine corrections. With the Gaussian approximation, the Sonine polynomials are not considered.
4. *Large mass of the Intruder*: within this particular limit, it is possible to take the first two terms of the Kramers-Moyal expansion, obtaining a diffusion process.

After this procedure, a simple scenario for the motion of the tracer emerges by observing Eq. (4.24). The effect of the other particles is translated in a effective drag and temperature, given by the simultaneous action of the bath and the surrounding particles. It is important to stress that, after this procedure, every information concerning the non equilibrium peculiarities of the system seems to be disappeared. Namely, given Eq. (4.24), the equilibrium fluctuation theorem is satisfied with respect to the "effective" temperature T_{tr} and no violations can occur. Moreover, the time reversal symmetry (i.e. the detailed balance) is satisfied. These considerations seems to be in strong contrast with the presence of dissipative forces associated to inelastic collisions. This paradox can be easily solved if one thinks again on the content of information required in order to appreciate the non equilibrium behavior of a system. Let us write the collision rule between the tracer and a particle from the gas

$$\boldsymbol{V}' = \boldsymbol{V} - (1+\alpha)\frac{m}{M+m}[(\boldsymbol{V}-\boldsymbol{v})\cdot\hat{\boldsymbol{\sigma}}]\hat{\boldsymbol{\sigma}}, \tag{4.29}$$

with the obvious meaning of the symbols. It is easy to see that, from the tracer "point of view", the collision is equivalent an elastic one with an effective mass $M' = 2\frac{M+m}{1+\alpha} - m \simeq \frac{2}{1+\alpha}M$ (for large mass).

So, if we observe only particle 1 and we do not know anything about the other particle involved in the collision, we cannot conclude that the interaction among the particles preserves or not the conservation of energy. This sort of "indistinguishability" is well reproduced in Eq. (4.24). Because of the absence of memory effects, the noise induced by the surrounding particles acts on the same time scale of the external noise with the result of being indistinguishable one from the other.

In conclusion, this is an interesting example of how a reduction of degrees of freedom, or a sort of projection mechanism can produce a lack of information, returning the "false" idea that the system is in equilibrium.

4.2 The Dense Case and the Effect of the Recollisions: A Numerical Study

The general lesson we learn from the previous section can be summarized with two main sentences:

- If one assumes molecular chaos and, consequently, absence of memory terms, one cannot have non-equilibrium effects.
- Within the Gaussian approximation scheme, in order to have a proper measure of irreversibility, the observation of a single degree of freedom is not sufficient.

In order to find a breakdown of the above mechanisms, it appears natural to study the dense inelastic case, where one expects that the molecular chaos condition is no more satisfied because of recollision effects and pure non-equilibrium conditions, like equilibrium fluctuation-dissipation violations and entropy production must be

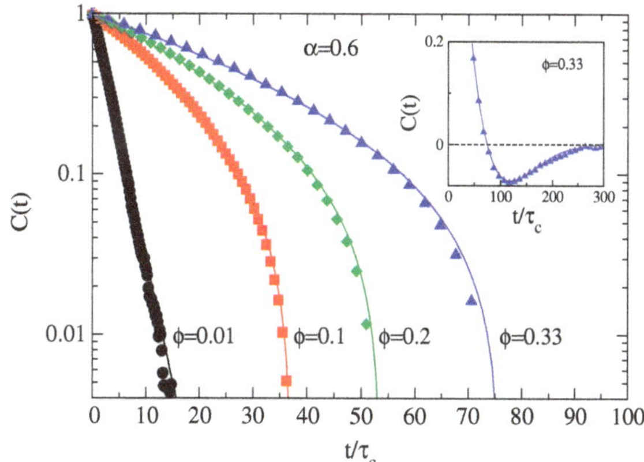

Fig. 4.1 Semi-log plot of $C(t)$ (*symbols*) for different values of $\phi = 0.01, 0.1, 0.2, 0.33$ at $\alpha = 0.6$. Times are rescaled by the mean collision time τ_c. Continuous *lines* are the best fits obtained with Eq. (4.33). *Inset* $C(t)$ and the best fit in linear scale for $\phi = 0.33$ and $\alpha = 0.6$

observed. In this regime, a proper analytically solvable approach is lacking, therefore the dilute theory developed in the previous section will be useful to have a well defined "equilibrium" limit.

As the packing fraction is increased, a numerical measure of the velocity autocorrelation $C(t) = \langle V(t)V(0)\rangle/\langle V^2\rangle$ and its response function $R(t) = \overline{\delta V(t)}/\delta V(0)$ (i.e. the mean response at time t to an impulsive perturbation applied at time 0) show an exponential decay modulated in amplitude by oscillating functions, not predictable from a simple Langevin equation [14]. Moreover violations of the Einstein relation $C(t) = R(t)$ are observed for $\alpha < 1$ [12, 15].

Molecular dynamics simulations of the system have been performed, giving access to $C(t)$ and $R(t)$, for several different values of the parameters α and ϕ. In Fig. 4.1, symbols correspond to the velocity correlation functions measured in the inelastic case, $\alpha = 0.6$, for different values of the packing fraction ϕ. The other parameters are fixed: $N = 2,500$, $m = 1$, $M = 25$, $r = 0.005$, $R = 0.025$, $T_b = 1$, $\gamma_b = 200$. Times are rescaled by the mean collision times τ_c, as measured in the different cases. Numerical data are averaged over $\sim 10^5$ realizations.

Notice that the Enskog approximation (4.6) cannot predict the observed functional forms, because it only modifies by a constant factor the collision frequency. In order to describe the full phenomenology, a model with more than one characteristic time is needed. The proposed model is a Langevin equation with a single exponential memory kernel

$$M\dot{V}(t) = -\int_{-\infty}^{t} dt'\, \Gamma(t-t')V(t') + \mathcal{E}'(t), \tag{4.30}$$

where, in this case,

$$\Gamma(t) = 2\gamma_0\delta(t) + \gamma_1/\tau_1 e^{-t/\tau_1} \tag{4.31}$$

and $\mathcal{E}'(t) = \mathcal{E}_0(t) + \mathcal{E}_1(t)$, with

$$\langle \mathcal{E}_0(t)\mathcal{E}_0(t')\rangle = 2T_0\gamma_0\delta(t-t'),\ \langle \mathcal{E}_1(t)\mathcal{E}_1(t')\rangle = T_1\gamma_1/\tau_1 e^{-(t-t')/\tau_1}, \tag{4.32}$$

the limit $\alpha \to 1$, the parameter T_1 is meant to tend to T_0 in order to fulfill the fluctuation dissipation relation of the 2nd kind $\langle \mathcal{E}'(t)\mathcal{E}'(t')\rangle = T_0\Gamma(t-t')$. Within this model the dilute case is recovered if $\gamma_1 \to 0$. In this limit, the parameters γ_0 and T_0 coincide with Γ_E and T_{tr} of the dilute theory (4.24).

The model (4.30), as shown in Sect. 3.2.2, predicts $C = f_C(t)$ and $R = f_R(t)$ with

$$f_{C(R)} = e^{-gt}[\cos(\omega t) + a_{C(R)}\sin(\omega t)]. \tag{4.33}$$

The variables g, ω, a_C and a_R are known algebraic functions of γ_0, T_0, γ_1, τ_1 and T_1. In particular, the ratio $a_C/a_R = [T_0 - \Omega(T_1 - T_0)]/[T_0 + \Omega(T_1 - T_0)]$, with $\Omega = \gamma_1/[(\gamma_0 + \gamma_1)(\gamma_0/M\tau_1 - 1)]$. Hence, in the elastic ($T_1 \to T_0$) as well as in the dilute limit ($\gamma_1 \to 0$), one gets $a_C = a_R$ and recovers the fluctuation dissipation relation $C(t) = R(t)$.

In Fig. 4.1 the continuous lines show the result of the best fits obtained using Eq. (4.33) for the correlation function, at restitution coefficient $\alpha = 0.6$ and for different values of the packing fraction ϕ. The functional form fits very well the numerical data. A multi-branch fit of measured C and R against Eq. (4.33), together with a measure of $\langle V^2\rangle$, yields five independent equations to determine the five parameters entering the model. We used the external parameters mentioned before, changing α or the box area A (to change ϕ) or the intruder's radius R, in order to change ϕ keeping or not keeping constant $\gamma_g \sim 1/l_0 \to 0$ (indeed different dilute limits can be obtained, where collisions matter or not).

Looking for an insight of the relevant physical mechanisms underlying the memory effect and in order to make clear the meaning of the parameters, it is useful to map Eq. (4.30) onto a Markovian equivalent model by introducing an auxiliary field (see Sect. 3.2):

$$\begin{aligned} M\dot{V} &= -\gamma_0(V - U) + \sqrt{2\gamma_0 T_0}\mathcal{E}_V \\ M'\dot{U} &= -\Gamma' U - \gamma_0 V + \sqrt{2\Gamma' T_1}\mathcal{E}_U, \end{aligned} \tag{4.34}$$

where $\mathcal{E}_V$ and $\mathcal{E}_U$ are white noises of unitary variance while $\Gamma' = \frac{\gamma_0^2}{\gamma_1}$ and $M' = \frac{\gamma_0^2\tau_1}{\gamma_1}$ are. In the chosen form (4.35), the dynamics of the tracer is remarkably simple: indeed V follows a memoryless Langevin equation in a *Lagrangian frame* with respect to a local field U, which is the *local average velocity field* of the gas particles colliding with the tracer. Extrapolating such an identification to higher densities, we are able to understand the value for most of the parameters of the model:

- The self drag coefficient of the intruder in principle is not affected by the change of reference to the Lagrangian frame, so that one predicts $\gamma_0 \sim \Gamma_E$
- For the same reason $T_0 \sim T_{tr}$

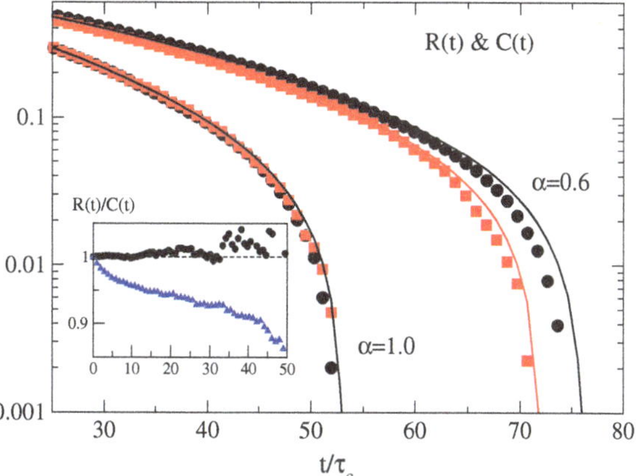

Fig. 4.2 *Left* correlation function $C(t)$ (*black circles*) and response function $R(t)$ (*red squares*) for $\alpha = 1$ and $\alpha = 0.6$, at $\phi = 0.33$. Continuous *lines* show the best fits curves obtained with Eq. (4.33). *Inset* the ratio $R(t)/C(t)$ is reported in the same cases

- T_1 is the "temperature" of the local field U. One expects that $T_1 \sim T_b$, since, thanks to momentum conservation, inelasticity does not affect the average velocity of a group of particles which almost only collide with themselves.

Fits from numerical simulations agree with this predictions (see [16] for details). Moreover it is also interesting to notice that at high density $T_{tr} \sim T_g \sim T_g^E$, which is probably due to the stronger correlations among particles. Finally we notice that, at large ϕ, $T_{tr} > T_{tr}^E$, which is coherent with the idea that correlated collisions dissipate *less* energy.

4.2.1 The Response Analysis and the Average Velocity Field

The system of coupled Langevin equations in (4.35) is able to reproduce the violations of the fluctuation dissipation relation measured in numerical simulations, as show in Fig. 4.2: symbols correspond to numerical data and continuous lines to the best fit curves, and in the inset the ratio $R(t)/C(t)$ is also reported. As largely discussed in Chap. 3, a relation between the response and correlations measured in the unperturbed system still exists, but, in the non-equilibrium case, one must take into account the contribution of the cross correlation $\langle V(t)U(0)\rangle$, i.e.:

$$R(t) = aC(t) + b\langle V(t)U(0)\rangle, \tag{4.35}$$

with $a = [1 - \gamma_1/M(T_0 - T_1)\Omega_a]$ and $b = (T_0 - T_1)\Omega_b$, where Ω_a and Ω_b are known functions of the parameters. At equilibrium, where $T_0 = T_1$, the fluctuation dissipation relation is recovered.

The mapping in a two variable system expressed in (4.35) is not only a mathematical trick, but produces strong differences for some observables like the entropy production, as extensively discussed in Chap. 3. The results given in Sect. 4.2.1 are consistent with a local velocity field, coupled with the dynamics of the tracer. Therefore, it is necessary to have an operative definition of this field, which can be exploited numerically. The field U is numerically defined as the average of particles velocities in the neighborhood $\mathcal{B}$ of the tracer, $U_{\mathcal{B}} = 1/N_{\mathcal{B}} \sum_{i \in \mathcal{B}} v_i$, where $N_{\mathcal{B}}$, and a proper choice of the size of $\mathcal{B}$ is commented in [16].

It is worth to notice that the correlation between the velocity of the tracer and the so-defined average velocity field is a pure non-equilibrium effect, emerging as soon as dissipative forces, like inelastic collisions, are present and which in not present in an elastic fluid, although dense.

4.2.2 Numerical Verification of the Gallavotti-Cohen Theorem

An important independent assessment of the effectiveness of model (4.35) comes from the study of the fluctuating entropy production. Given the trajectory in the time interval $[0, t]$, $\{V(s)\}_0^t$, and its time-reversed $\{\mathcal{I}V(s)\}_0^t \equiv \{-V(t-s)\}_0^t$, in Sect. 3.2.3 it has been shown that the entropy production for the model (4.35) takes the form (3.60) that we rewrite here for simplicity:

$$\Sigma_t = \log \frac{P(\{V(s)\}_0^t)}{P(\{\mathcal{I}V(s)\}_0^t)} \approx \gamma_0 \left(\frac{1}{T_0} - \frac{1}{T_1} \right) \int_0^t ds \; V(s)U(s). \tag{4.36}$$

Boundary terms, in the stationary state, are subleading for large t and have been neglected. This functional vanishes exactly in the elastic case, $\alpha = 1$, where equipartition holds, $T_1 = T_0$, and is zero on average in the dilute limit, where $\langle VU \rangle = 0$. Formula (4.36) reveals that the leading source of entropy production is the energy transferred by the "force" $\gamma_0 U$ on the tracer, weighed by the difference between the inverse temperatures of the two "thermostats".

Then, measuring the entropy production of Eq. (4.36) (by replacing $U(t)$ with $U_{\mathcal{B}}$) along many trajectories of length t, we can compute the probability $P(\Sigma_t = x)$ and compare it to $P(\Sigma_t = -x)$, in order to verify the Fluctuation Relation

$$\log \frac{P(\Sigma_t = x)}{P(\Sigma_t = -x)} = x. \tag{4.37}$$

In Fig. 4.3 we report our numerical results. The main frame confirms that at large times the Fluctuation Relation (4.37) is well verified. The inset shows the collapse of $\log P(\Sigma_t)/t$ onto the large deviation rate function for large times.

In conclusion, we can say that a remarkable test of the model (4.35) is in its ability to allow for a perfect recovery of the fluctuation theorem, together with the prediction of the out of equilibrium response. Indeed it must be noticed that formula (4.36) does not contain further parameters but the ones already determined by correlation and response measure, namely the slope of the graph is not adjusted by further fits. As a consequence, a wrong evaluation of the weighing factor $(1/T_0 - 1/T_1) \approx$

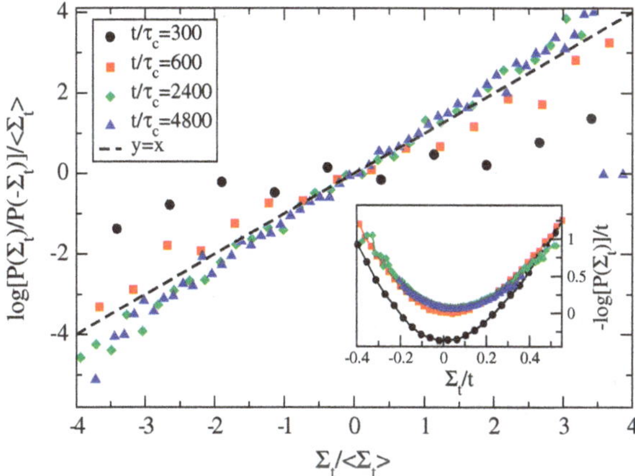

Fig. 4.3 Check of the fluctuation relation (4.37) in the system with $\alpha = 0.6$ and $\phi = 0.33$. Inset: collapse of the rescaled probability distributions of Σ_t at large times onto the large deviation function

$(1/T_{tr} - 1/T_b)$ or of the "energy injection rate" $\gamma_0 U(t)V(t)$ in Eq. (4.36) could produce a completely different slope in Fig. 4.3, suggesting a "violation" of the theorem.

This is a remarkable progress if compared with the previous result on entropy production in granular gases, as discussed in Sect. 2.3.3. Within this theory the main ingredient is given by the emerging coupling between the tracer and the velocity field. It must be stressed that several non-linear effects of the system have been neglected in this description and this is possible, as explained in Sect. 4.1.2.1, thanks to the scale separation limit. In this direction, the verification of the fluctuation theorem it is a strong signal that a main source of entropy production has been well modeled.

References

1. Williams, D.R.M., MacKintosh, F.C.: Driven granular media in one dimension: correlations and equation of state. Phys. Rev. E **54**, R9 (1996)
2. van Noije, T.P.C., Ernst, M.H., Trizac, E., Pagonabarraga, I.: Randomly driven granular fluids: large-scale structure. Phys. Rev. E **59**, 4326 (1999)
3. Puglisi, A., Loreto, V., Marconi, U.M.B., Petri, A., Vulpiani, A.: Clustering and non-gaussian behavior in granular matter. Phys. Rev. Lett. **81**, 3848 (1998)
4. Puglisi, A., Visco, P., Trizac, E., van Wijland, F.: Dynamics of a tracer granular particle as a nonequilibrium Markov process. Phys. Rev. E **73**(2), 021301 (2006)
5. Brilliantov, N.K., Poschel, T.: Kinetic Theory of Granular Gases. Oxford University Press, Oxford (2004)
6. Chapman, S., Cowling, T.: The mathematical theory of non-uniform gases, vol 1. Cambridge University Press, London (1991)
7. Pöschel, T., Brilliantov, N. (eds.): Granular Gas Dynamics. Lecture Notes in Physics, vol 624. Springer, Berlin (2003)

8. Risken, H.: The Fokker-Planck equation: Methods of solution and applications. Springer, Berlin (1989)
9. Kubo, R., Toda, R.: Statistical physics II: nonequilibrium stastical mechanics. Springer, Berlin (1991)
10. Marconi, U.M.B., Puglisi, A., Rondoni, L., Vulpiani, A.: Fluctuation-dissipation: response theory in statistical physics. Phys. Rep. **461**, 111 (2008)
11. Sarracino, A., Villamaina, D., Costantini, G., Puglisi, A.: Granular brownian motion. J. Stat. Mech. (2010). doi:10.1088/1742-5468/2010/04/P04013
12. Puglisi, A., Baldassarri, A., Vulpiani, A.: Violations of the Einstein relation in granular fluids: the role of correlations. J. Stat. Mech. (2007). doi : 10.1088/1742-5468/2007/08/P08016
13. Bunin, G., Shokef, Y., Levine, D.: Frequency-dependent fluctuation-dissipation relations in granular gases. Phys. Rev. E **77**, 051301 (2008)
14. Fiege, A., Aspelmeier, T., Zippelius, A.: Long-time tails and cage effect in driven granular fluids. Phys. Rev. Lett. **102**, 098001 (2009)
15. Villamaina, D., Puglisi, A., Vulpiani, A.: The fluctuation-dissipation relation in sub-diffusive systems: the case of granular single-file diffusion. J. Stat. Mech. (2008). doi:10.1088/1742-5468/2008/10/L10001
16. Sarracino, A., Villamaina, D., Gradenigo, G., Puglisi, A.: Irreversible dynamics of a massive intruder in dense granular fluids. Europhys. Lett. **92**, 34001 (2010)

Chapter 5
Anomalous Transport and Non-Equilibrium

Abstract This chapter the additional ingredient of anomalous diffusion, combined with non equilibrium conditions, is studied. The analysis starts with a random walk on a comb lattice, which can be analytically solvable. A detailed analysis of the "single file model" is then shown and the response analysis is similar to that one in higher dimensions. The chapter ends with the study of a ratchet effect in a fragile glass former. Because of the presence of disorder, under certain conditions, an intruder in a glass former exhibits subdiffusion. Despite of the great differences from the "family" of the non equilibrium steady states, also in this system it is possible to observe a ratchet effect, although characterized by a subvelocity due to the disorder.

A large variety of systems exhibits normal diffusion, which is given by the linear growing in time of the mean square displacement of a tracer particle. Such a behavior is strictly connected with the central limit theorem. As a consequence, when the conditions of the central limit theorem are not satisfied, some systems can show the so-called anomalous diffusion.

A natural issue is whether the presence of anomalous diffusion can significantly change the "response scenario", in both equilibrium and non-equilibrium setups. In order to blaze an answer to this question, in the first part of the chapter, two subdiffusdive models are analyzed by means of the generalized response relation. Because of the presence of non-trivial couplings and memory effects, the equilibrium response is interesting on its own. First, a one-dimensional random walk on a comb lattice is presented, where the trapping mechanism is put under analysis, and the knowledge of the transition rates makes possible an analytical approach. The more realistic single file model is then studied, where the non-overlapping properties of the particles originates memory terms and subdiffusive transport properties.

The second part is devoted to the study of the ratchet effect in an aging glass former. Such a study is interesting for two main reasons: first, the glass former is not able to equilibrate and exhibit aging: therefore there is a big difference respect to the rest of the models studied in this work. Secondly, anomalous transport isinduced by

D. Villamaina, *Transport Properties in Non-Equilibrium and Anomalous Systems*, Springer Theses, DOI: 10.1007/978-3-319-01772-3_5,

disorder when a quenching is performed well below the mode coupling temperature. So we have an unusual example of a ratchet effect in a non-stationary subdiffusive system.

5.1 The Problem of Anomalous Diffusion

The models discussed in the previous chapters have a behavior qualitatively similar to the Brownian motion, namely, for large times, the following relation holds:

$$\langle \delta x^2(t) \rangle \simeq 2Dt, \tag{5.1}$$

where D is the diffusion coefficient. By considering that $x(t) = \int_0^t v(t)$, if for v a steady state distribution probability can be defined, the diffusion coefficient is linked to the velocity autocorrelation via the Kubo formula

$$D \equiv \int_0^t \langle v(t)v(0) \rangle dt. \tag{5.2}$$

Note that the result (5.1) is, in a sense, a consequence of the central limit theorem, if it holds: the position of the particle x is nothing but the sum of its velocity on time, and, therefore the statistical properties of velocity are determinant. In particular:

- if the variance of the velocity $\langle v^2 \rangle$ is finite,
- if the velocity autocorrelation is fast decaying in time,

the central limit theorem is valid and normal diffusion takes place. On the other hand, when at least one of the two conditions above is not valid, anomalous transport properties can emerge, characterized by a scaling behavior of the kind:

$$\langle \delta x^2(t) \rangle \propto t^{2\nu}, \tag{5.3}$$

with $\nu \neq \frac{1}{2}$. In particular, the system is defined to be superdiffusive when $\nu > \frac{1}{2}$ ($D = \infty$) and, subdiffusive when $\nu < \frac{1}{2}$ ($D = 0$). Anomalous diffusion is present in a huge variety of natural phenomena and different models exhibiting subdiffusion are studied in literature (for reviews see [8, 28]).

One may wonder if the knowledge of the exponent ν in (5.3) is sufficient to determine the distribution properties of the system under investigation. Roughly speaking, a natural way to proceed is to assume a simple scaling of the higher moments $\overline{x^q} \sim t^{q\nu}$, which can be translated into

$$P(x,t) \sim e^{F(x/t^{\nu})}. \tag{5.4}$$

Moreover, one can assume a sort of "generalized" large deviation hypothesis

$$P(x,t) \sim e^{tf(x/t)}. \tag{5.5}$$

In conclusion, a direct comparison between (5.4) and (5.5) yields to the following identification

$$F(x/t^{\nu}) = tf(x/t). \tag{5.6}$$

Deriving both the sides of (5.6) with respect to x and with respect to t and putting together the obtained relations, one has the following differential equation for $F(x/t^{\nu})$

$$(1-\nu)\frac{x}{t^{\nu}}F'(x/t^{\nu}) = F(x/t^{\nu}), \tag{5.7}$$

the solution of (5.7) leads to the following distribution function

$$P(x,t) \sim e^{-c(x/t^{\nu})^{\alpha}}, \tag{5.8}$$

where $\alpha = \frac{1}{1-\nu}$ and c is a constant depending on the parameters. The result expressed in (5.8) has been originally proposed by Fisher for polymers [15] and it is numerically verified in a class of models like, for instance, the random walk of a comb lattice, as shown in Sect. 5.2, but it is not verified for all the models exhibiting anomalous diffusion. Indeed, there are several models for which (5.5) is not valid, for different reasons:

- The condition (5.4) is not valid in the case of strong anomalous diffusion, for which a dependence from q of the scaling exponent must be taken into account [11]
- The condition (5.5) is not always valid because of the presence of long time tails in the correlations of velocities. For instance, in the single-file model, Eq. (5.5) is not verified, as shown in Sect. 5.3.

Moreover, there is another class of models, like the Richardson diffusion [33], that leads to different relations between ν and α (cfr. Fig. 5.1). In conclusion, from the scaling of the second moment it is not possible to determine uniquely the probability distribution of positions $P(x,t)$, as shown in Fig. 5.1.

Starting from the above considerations, a new question can be posed: what happens when a perturbation is applied to an anomalous system? In order to give an answer to this issue, it is necessary to distinguish between subdiffusion and superdiffusion. In superdiffusive models, generally, the anomalous transport condition is linked with a certain value of a control parameter and can disappear for large times also for a small perturbation: this is the case, for instance, of some deterministic chaotic systems [6, 19, 37]. These aspects makes the problem quite hard and, also if recently some progresses have been done, the question is still open and challenging.

On the contrary, regarding subdiffusion, a recent result has been obtained on the so called fractional Fokker Planck equation [28].

The fractional-Fokker-Planck equation is given by

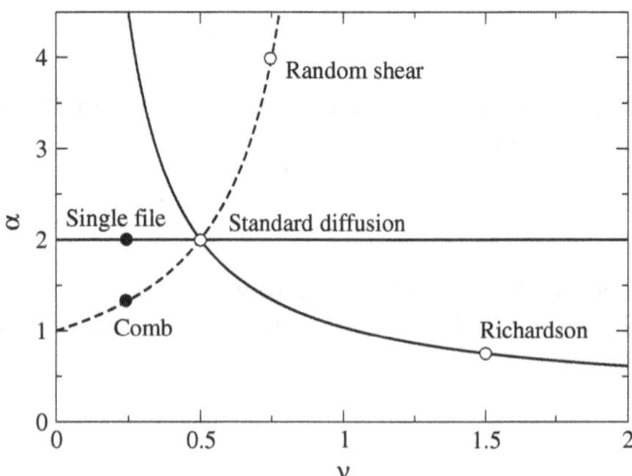

Fig. 5.1 Scheme of the relation between ν and α in different models. The full *circles* corresponds to the ones analyzed in this work. This figure is based on [25]

$$\frac{\partial}{\partial t}P(x,t) = \mathcal{D}_T^{1-\gamma}\left[\mathcal{L}_{FP}P(x,t)\right] \tag{5.9}$$

where $\mathcal{L}_{FP}$ is the usual Fokker-Planck operator and $\mathcal{D}_T^{1-\gamma}$ is the so-called Riemann-Liouville fractional operator defined through:

$$\mathcal{D}_T^{1-\gamma}\left[f(x,t)\right] = \frac{1}{\Gamma(\gamma)}\frac{\partial}{\partial t}\int_0^t \frac{f(x,t')}{(t-t')^{1-\gamma}}dt' \text{ with } 0 < \gamma \leq 1 \ , \tag{5.10}$$

being $\Gamma(\gamma)$ the Euler gamma function. In absence of an external potential, one has that $\langle x^2(t)\rangle \propto t^{\gamma}$, namely it exhibits subdiffusion. It has been shown [27] that if a constant force is applied (if one considers a potential of the kind $V = Fx$ in the $\mathcal{L}_{FP}$ operator), a generalized Einstein relation for the subdiffusive case holds

$$\overline{x(t)} = \frac{1}{2}\frac{F\langle x^2(t)\rangle}{k_B T} \ , \tag{5.11}$$

The investigation of charge carriers transport in semiconductors [18] showed that, up to a prefactor Eq. (5.11) is indeed valid.

The aim of this chapter is to go beyond this result, by analyzing the non-equilibrium response in two subdiffusive models, which are not fully described by the Fractional Fokker Planck Equation.

5.2 The Random Walk on a Comb Lattice

The comb lattice is a discrete structure (for a list of references, see [32]) consisting of an infinite linear chain (backbone), the sites of which are connected with other linear chains (teeth) of length L (cfr. Fig. 5.2). We denote by $x \in (-\infty, \infty)$ the position of

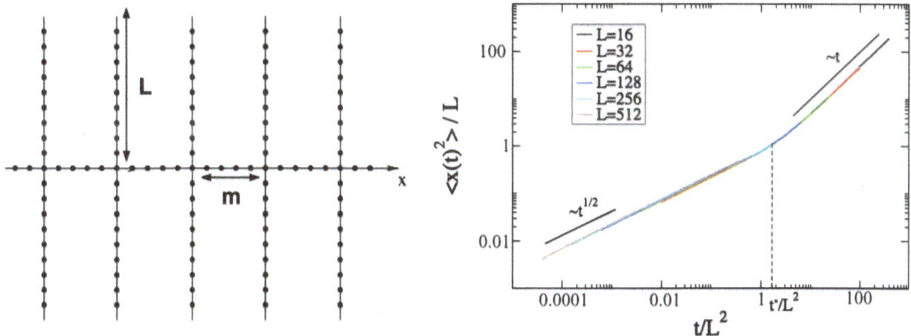

Fig. 5.2 *Left panel* Schematic representation of the Comb lattice. In our case, we always set the parameter m equal to one. *Right panel* $\langle x^2(t)\rangle_0/L$ *vesus* t/L^2 is plotted for several values of L in the comb model, and the data collapse is shown

the particle performing the random walk along the backbone and with $y \in [-L, L]$ that along a tooth. The discrete time transition probabilities from (x, y) to (x', y') are:

$$\begin{aligned} W^d[(x,0) \to (x\pm 1,0)] &= 1/4 \pm d \\ W^d[(x,0) \to (x,\pm 1)] &= 1/4 \\ W^d[(x,y) \to (x,y\pm 1)] &= 1/2 \quad \text{for} \quad y \neq 0, \pm L. \end{aligned} \tag{5.12}$$

On the boundaries of each tooth, namely when $y = \pm L$, the particle is reflected with probability 1. Of course, the normalization $\sum_{(x',y')} W^d[(x,y) \to (x',y')] = 1$ holds. The case $L = \infty$ is obtained in numerical simulations by letting the y coordinate increase without boundaries.

Let us first describe the diffusion properties in the case $d = 0$. For finite teeth length $L < \infty$, we have numerical evidence of a dynamical crossover from a subdiffusive to a simple diffusive asymptotic behavior[1] (see Fig. 5.2)

$$\langle x^2(t)\rangle_0 \simeq \begin{cases} Ct^{1/2} & t < t^*(L) \\ 2D(L)t & t > t^*(L), \end{cases}$$

where C is a constant and $D(L)$ is an effective diffusion coefficient depending on L. We find $t^*(L) \sim L^2$ and $D(L) \sim 1/L$, as shown in Fig. 5.2.

In the limit of infinite teeth, $L \to \infty$, $D \to 0$ and $t^* \to \infty$ and the system shows subdiffusive behavior

$$\langle x^2(t)\rangle_0 \sim t^{1/2}. \tag{5.13}$$

[1] The symbol $\langle \ldots \rangle_0$ denotes an average over different realizations of the dynamics (5.12) with $d = 0$ and initial condition $x(0) = y(0) = 0$.

In this case, the probability distribution function behaves as

$$P_0(x,t) \sim t^{-1/4} e^{-c\left(\frac{|x|}{t^{1/4}}\right)^{4/3}}, \tag{5.14}$$

the behavior (5.14) also holds in the case of finite L, provided that $t < t^*$. For larger times a Gaussian distribution is observed. As mentioned above, this is an example in which the *ansatz* (5.8) is largely verified.

These numerical observations are clarified if one consider the trapping effect induced by the presence of the teeth. It is possible to calculate the probability that the walker reemerges at time τ in a finite tooth of length L: it decays as $\tau^{-3/2}$ for $\tau < L^2$ and goes down exponentially fast for longer times [39]. Then the average time spent in a tooth is

$$\langle\tau\rangle \sim \int^{+\infty} d\tau \tau \tau^{-3/2} \sim L \sim \int^{L^2} d\tau \tau^{-1/2} \sim L, \tag{5.15}$$

and diverges with L that goes with infinity. From this results we learn that:

- When L goes to infinity, one has a broad distribution for the returning time, which decays as $t^{-3/2}$. In this limit, it is possible to show that the diffusion on the backbone is equivalent to a Continuous time random walk of parameter $\mu = \frac{1}{2}$ [32] and the subdiffusion is explained.
- In the case of $L < \infty$, for times lower than L^2, the walker does not feel the finiteness of the teeth, and the system show subdiffusion. On the contrary, for larger times normal diffusion is recovered, with an effective diffusion coefficient $D_{eff} \sim \frac{1}{\langle\tau\rangle} \sim \frac{1}{L}$.

5.2.1 Response Properties

In the following we present the numerical observation obtained by studying the response on a comb structure.

The perturbation induced mimics an electric field acting on the particle: it unbalances to the right the transition rates connecting two sites on the backbone, namely the perturbed transition rates are

$$W^{d+\epsilon}[(x,0) \to (x \pm 1, 0)] = \frac{1}{4} + d \pm \epsilon, \tag{5.16}$$

we stress again that the perturbation does not act on sites of the teeth. By pursuing the analogy with the electric field, it is clear that there are two relevant physical conditions:

- $d = 0$: the unperturbed process is given by a symmetric random walk on a comb structure. Moreover, in the case $L < \infty$ apart from the initial anomalous transient, the position distribution is Gaussian and standard diffusion takes place. This case, for our purposes, can be considered equivalent to an equilibrium case.
- $d \neq 0$: a current is flowing into the system and drives it out of equilibrium. Then, the detailed balance breaking is combined with the subdiffusive properties.

As a consequence, according to these considerations, the response behavior in these two cases is expected to be qualitatively different.

The first relevant consideration which emerges from numerical analysis is that the fluctuation dissipation relation in its standard form is fulfilled, namely one has evidence that

$$\langle x^2(t)\rangle_0 \simeq C\overline{\delta x}(t) \sim t^{1/2}. \tag{5.17}$$

Moreover, the proportionality between $\langle x^2(t)\rangle_0$ and $\overline{\delta x}(t)$ is fulfilled also with $L < \infty$, where both the mean square displacement and the drift with an applied force exhibit the same crossover from subdiffusive, $\sim t^{1/2}$, to diffusive, $\sim t$. Therefore what we can say is that the fluctuation dissipation relation is somehow "blind" to the dynamical crossover experienced by the system. When the perturbation is applied to a state without any current, the proportionality between response and correlation holds despite anomalous transport phenomena.

On the contrary, at variance with the depicted above scenario about the zero current situation, within a state with a non zero drift [10] the emergence of a dynamical crossover is connected to the breaking of the FDR. Indeed, the mean square displacement in the presence of a non zero current, even with $L = \infty$, is

$$\langle x^2(t)\rangle_d \sim at^{1/2} + bt, \tag{5.18}$$

where a and b are two constants, whereas

$$\overline{\delta x_d}(t) \sim t^{1/2}, \tag{5.19}$$

with $\overline{\delta x_d}(t) = \langle x(t)\rangle_{d+\varepsilon} - \langle x(t)\rangle_d$: at large times the Einstein relation breaks down (see Fig. 5.3). The proportionality between response and fluctuations cannot be recovered by simply replacing $\langle x^2(t)\rangle_d$ with $\langle x^2(t)\rangle_d - \langle x(t)\rangle_d^2$, as it happens for Gaussian processes, namely we find numerically

$$\langle [x(t) - \langle x(t)\rangle_d]^2\rangle_d \sim a't^{1/2} + b't, \tag{5.20}$$

where a' and b' are two constants, as reported in Fig. 5.3.

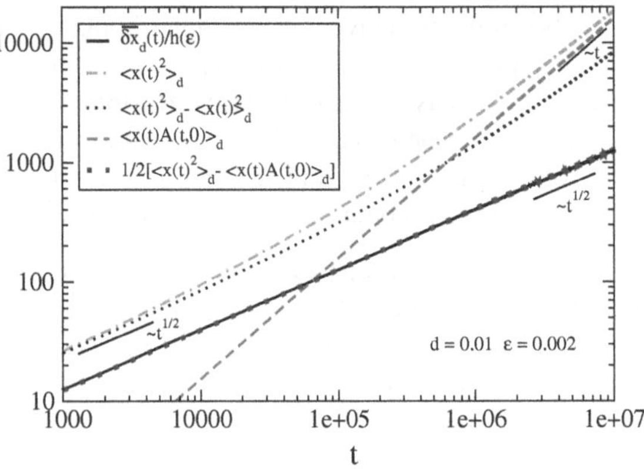

Fig. 5.3 Response function (*black line*), Mean square displacement (*red dotted line*) and second cumulant (*black dotted line*) measured in the the comb model with $L = \infty$, field $d = 0.01$ and perturbation $\varepsilon = 0.002$. The correlation with $A(t, 0)$ (*green dotted line*) yields the right correction to recover the full response function (*blue dotted line*), in agreement with the fluctuation dissipation relation (5.24)

5.2.1.1 Application of the Generalized Response Formula

The results of the previous section show that the first moment of the probability distribution function with drift $P_d(x, t)$ and the second cumulant of $P_0(x, t)$ are always proportional. Differently, the first cumulant of $P_{d+\varepsilon}(x, t)$ is not proportional to the second moment of $P_d(x, t)$, namely $\langle x(t)\rangle_{d+\varepsilon} \not\sim \langle x^2(t)\rangle_d - \langle x(t)\rangle_d^2$. In order to find out a relation between such quantities, we need to use a generalized fluctuation-dissipation relation of the form (2.25).

According with the definitions of the transition rates, one has for the backbone

$$W^{d+\varepsilon} = W^d[(x, y) \to (x', y')]\left(1 + \frac{\varepsilon(x' - x)}{W^0 + d(x' - x)}\right) \simeq W^d e^{\frac{\varepsilon}{W^0}(x'-x)}, \tag{5.21}$$

where $W^0 = 1/4$, and the last expression holds under the condition $d/W^0 \ll 1$. Regarding the above expression as a "local detailed balance" condition for our Markov process can be rewritten, for $(x, y) \neq (x', y')$, as

$$W^{d+\varepsilon}[(x, y) \to (x', y')] = W^d[(x, y) \to (x', y')]e^{\frac{h(\varepsilon)}{2}(x'-x)}, \tag{5.22}$$

where $h(\varepsilon) = 2\varepsilon/W^0$. As a consequence, from (2.25) follows

$$\frac{\overline{\delta \mathcal{O}_d}}{h(\varepsilon)} = \frac{\langle \mathcal{O}(t)\rangle_{d+\varepsilon} - \langle \mathcal{O}(t)\rangle_d}{h(\varepsilon)} \tag{5.23}$$

$$= \frac{1}{2}\left[\langle \mathcal{O}(t)x(t)\rangle_d - \langle \mathcal{O}(t)x(0)\rangle_d - \langle \mathcal{O}(t)A(t, 0)\rangle_d\right], \tag{5.24}$$

where $\mathcal{O}$ is a generic observable, and $A(t, 0) = \sum_{t'=0}^{t} B(t')$, with

$$B[(x, y)] = \sum_{(x',y')} (x' - x)W^d[(x, y) \to (x', y')]. \quad (5.25)$$

Recalling the definitions (5.12), from the above equation we have

$$B[(x, y)] = 2d\delta_{y,0}. \quad (5.26)$$

The numerical results described in the previous section can be then read in the light of the fluctuation-dissipation relation (5.24):

- Considering $\mathcal{O}(t) = x(t)$, in the case without drift, i.e. $d = 0$, one has $B = 0$ and, recalling the choice of the initial condition $x(0) = 0$,

$$\frac{\overline{\delta x}}{h(\varepsilon)} = \frac{\langle x(t)\rangle_\varepsilon - \langle x(t)\rangle_0}{h(\varepsilon)} = \frac{1}{2}\langle x^2(t)\rangle_0. \quad (5.27)$$

  Equation (5.27) explains the observed behavior (5.17) even in the anomalous regime and predicts the correct proportionality factor, $\overline{\delta x}(t) = \varepsilon/W^0\langle x^2(t)\rangle_0$.
- Considering $\mathcal{O}(t) = x(t)$, in the case with $d \neq 0$, one has

$$\frac{\overline{\delta x_d}}{h(\varepsilon)} = \frac{1}{2}\left[\langle x^2(t)\rangle_d - \langle x(t)A(t, 0)\rangle_d\right]. \quad (5.28)$$

  Equation (5.28) explains the observed behaviors (5.18) and (5.19): the leading behavior at large times of $\langle x^2(t)\rangle_d \sim t$, turns out to be exactly canceled by the term $\langle x(t)A(t, 0)\rangle_d$, so that the relation between response and unperturbed correlation functions is recovered (see Fig. 5.3).
- As discussed above, it is not enough to substitute $\langle x^2(t)\rangle_d$ with $\langle x^2(t)\rangle_d - \langle x(t)\rangle_d^2$ to recover the proportionality with $\overline{\delta x_d}(t)$ when the process is not Gaussian. This can be explained in the following manner. By making use of the second order out of equilibrium fluctuation dissipation relation, derived by Lippiello et al. in [23], we can explicitly evaluate

$$\langle x^2(t)\rangle_d = \langle x^2(t)\rangle_0 + h^2(\epsilon)\frac{1}{2}\left[\frac{1}{4}\langle x^4(t)\rangle_0 + \frac{1}{4}\langle x^2(t)A^{(2)}(t, 0)\rangle_0\right], \quad (5.29)$$

  where $A^{(2)}(t, 0) = \sum_{t'=0}^{t} B^{(2)}(t')$ with

$$B^{(2)} = -\sum_{x'}(x' - x)^2 W[(x, y) \to (x', y')] = -1/2\delta_{y,0}. \quad (5.30)$$

  Then, recalling Eq. (5.27), we obtain

$$\langle \delta x^2(t) \rangle_d = \langle x^2(t) \rangle_d - \langle x(t) \rangle_d^2 = \langle x^2(t) \rangle_0 + h^2 \left[\frac{1}{8} \langle x^4(t) \rangle_0 + \frac{1}{8} \langle x^2(t) A^{(2)}(t,0) \rangle_0 - \frac{1}{4} \langle x^2(t) \rangle_0^2 \right]. \quad (5.31)$$

Numerical simulations show that the term in the square brackets grows like t yielding a scaling behavior with time consistent with Eq. (5.20).

5.2.1.2 Some Remarks

Let us underline some important points emerged from the use of the generalized response formulas. First of all, in absence of external currents, the subdiffusive version of the Einstein relation is recovered for the Comb case:

$$\overline{\delta x}(t) \propto \langle x^2(t) \rangle_0. \quad (5.32)$$

Note that, if one follows the derivation of this result, it is evident that the comb geometry is not the only one for which this result holds. In other words, many details of the shape of the teeth are not relevant (see Fig. 5.4). Another remarkable aspect is that Eq. (5.32) is blind to dynamical crossovers, which means a change in the shape of the $P(x,t)$.

Let us discuss the response in presence of the current. From Eq. (5.27) one realizes that, as soon as $d \neq 0$, a key role is ruled by the observable $A(t,0)$. The above observable yields an effective measure of the propensity of the system to leave a certain state (x, y) and, in some contexts, it is referred to as activity [3, 4]. Noticeably, in the case here analyzed, the sum on B has an intuitive meaning: it counts the time spent by the particle on the x axis, which is nothing but the region where the current is applied.

Let us conclude by underlining the role of non-Gaussianity. Let us consider a one-dimensional random walk on a simple line: in such a case, from (5.30), one has that $B^{(2)} = -1$ and $A^{(2)}(t,0) = -t$. This is straightforward derived if one considers

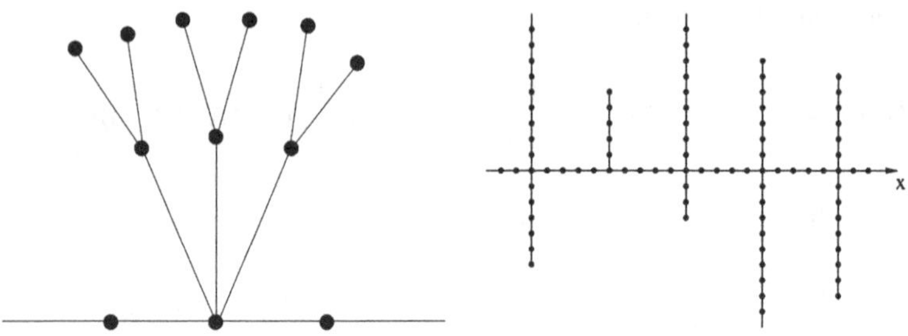

Fig. 5.4 Equation (5.32) is valid also for a more general class of structures, for instance in presence of threes (*left panel*) of disordered length of of the teeth

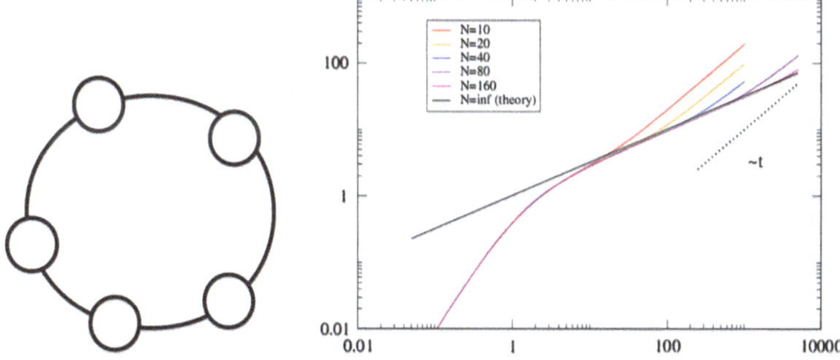

Fig. 5.5 *Left* Schematic picture of the single-file model for $N = 5$. *Right* Mean square displacement for a single particle for different N, denoting the double regime described in (5.3). The *black solid line* it the theoretical prediction for the case of infinite particles

that, in this case, the condition $\delta_{y,0}$ is always satisfied and $W(x \to x \pm 1) = \frac{1}{2} \pm d$. Then

$$\langle x^2(t)\rangle_d - \langle x(t)\rangle_d^2 = \langle x^2(t)\rangle_0 + h^2(\epsilon)\left[\frac{1}{8}\langle x^4(t)\rangle_0 - \frac{1}{8}t\langle x^2(t)\rangle_0 - \frac{1}{4}\langle x^2(t)\rangle_0^2\right]. \quad (5.33)$$

Since for a simple random walk, due to the Gaussian distribution, $\langle x^4(t)\rangle_0 = 3\langle x^2(t)\rangle_0^2$ and $\langle x^2(t)\rangle_0 = t$, the term in the square brackets vanishes identically and that explains why, in the presence of a drift, the second cumulant grows exactly as the second moment with no drift. In other words, in presence of currents, the *naive* correction

$$\overline{x(t)}_{d+\epsilon} - \overline{x(t)}_d = \langle x^2(t)\rangle_d - \langle x(t)\rangle_d^2 \quad (5.34)$$

is only possible in the Gaussian case (Fig 5.5). On the contrary, in case of non-Gaussian anomalous systems, it is no more valid.

5.3 The Single File Model

In the previous section we have studied a one particle system, whose subdiffusion is essentially due to a trapping mechanism. In this section, we will turn our attention on a many body system, known in literature with the name of "single-file" model. In this case, as we will see, the main mechanism responsible of the anomalous transport properties is given by the impossibility for the particles to overlap each other. The system given by N Brownian rods on a ring of length L interacting with elastic collisions and coupled with a thermal bath. The equation of motion for the i-th particle velocity between collisions is

$$m\dot{v}_i(t) = -\gamma v_i(t) + \eta_i(t), \tag{5.35}$$

where m is the mass, γ is the friction coefficient, and η is a white noise with variance $\langle \eta(t)\eta(t')\rangle = 2T\gamma\delta(t-t')$. The combined effect of collisions, noise and geometry (since the system is one-dimensional, the particles cannot overcome each other) produces a non-trivial behavior. In the thermodynamic limit, i.e. L, $N \to \infty$ with $N/L \to \rho$, a subdiffusive behavior occurs [20], we will see the details in the next section.

Analogously to the comb model, the case of N and L finite presents some interesting aspects. In order to avoid trivial results due to the periodic boundary conditions on the ring, it is suitable to define the position of a tagged particle as $s(t) = \int_0^t v(t')dt'$, where $v(t)$ is its velocity. For the mean square displacement $\langle s^2(t)\rangle$, averaged over the thermalized initial conditions and over the noise, we find, after a transient ballistic behavior for short times, a crossover between two different regimes:

$$\langle s^2(t)\rangle \simeq \begin{cases} \frac{2(1-\sigma\rho)}{\rho}\sqrt{\frac{D}{\pi}}t^{1/2} & t < \tau^*(N) \\ \frac{2D}{N}t & t > \tau^*(N), \end{cases}$$

where σ is the length of the rods and D is the diffusion coefficient of the single Brownian particle [22]. Note that the asymptotic behavior is completely determined by the motion of the center of mass. In a proper reference frame, indeed, it is possible to write the position of the single particle as the sum of two different contributions:

$$x_i = n(t) + \phi_i(t), \tag{5.36}$$

where $n(t)$ takes into account of the numbers of rounds completed and it is the same for all the particles and $\phi_i(t)$ is a bounded function between L and $-L$. Then, for large times, one has that[2]

$$\langle x_i^2(t)\rangle \simeq \langle n^2(t)\rangle \simeq \langle x^2(t)_{cm}\rangle. \tag{5.37}$$

In order to obtain the equation for the center of mass it is sufficient to sum over all the particles the equation (5.35) and divide by N, obtaining:

$$m\dot{v}_{cm}(t) = -\gamma v_{cm}(t) + \eta_{cm}(t) \tag{5.38}$$

$\langle \eta_{cm}(t)\eta_{cm}(t')\rangle = 2\gamma\frac{T}{N}\delta(t-t')$ which is not affected by the collisions and simply diffuses, with a coefficient $D_{eff} = \frac{D}{N}$, explaining the second part of the result (5.3). Moreover, as clear from numerical simulations, $\tau^* \sim N^2$ and in the limit of infinite number of particles the behavior becomes subdiffusive, in perfect analogy with what observed for the comb model, now the role of L is here played by N.

[2] Note that $\langle n(t)\phi_i(t)\rangle = 0$ for symmetry reasons.

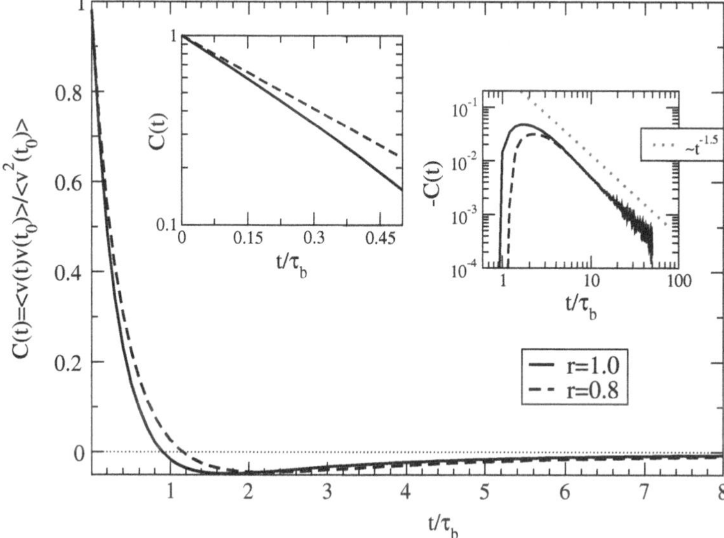

Fig. 5.6 Plot of the normalized autocorrelation $C(t)$ versus time, for two cases, one elastic (*full line*) and the other inelastic (*dashed line*). In the *left inset* we show a blow-up of the exponential decay at early times. In the *right inset* you can find a blow up in log-log scale of the negative tail, together with a power law decay $t^{-3/2}$. Here $\phi = 0.1$, and $\alpha \approx 0.9$

Analogously of what done in higher dimensions in Chap. 4, it is possible to introduce an inelastic version of this model.

In Fig. 5.6 we show the normalized autocorrelation function: $C(t) = C_v(t)/C_v(0) = \langle v(t)v(0)\rangle/T_g$ for the velocity of a tagged particle (a tracer with the same properties of other particles). In both elastic and inelastic experiments, $C(t)$ presents three main features:

- an exponential decay $C(t) \sim \exp(-t/\tau_{corr})$ at early times.
- a negative minimum at times of the order of the mean collision one.
- an asymptotic a power-law decay $C(t) \sim -t^{-3/2}$.

Note that the negative minimum is necessary to have subdiffusion, from relation (5.2) indeed, one has $\int_0^\infty C(t) = 0$, while the final power-law decay with 3/2 exponent is necessary to have $\langle x^2(t)\rangle \sim t^{1/2}$. The single-file diffusion scenario $\langle x^2(t)\rangle \sim t^{1/2}$ holds for any value of the parameters, as already reported in [12].

5.3.1 Equation of Motion and the Harmonization Procedure

As shown in Fig. 5.6, the velocity correlation function has a nontrivial behavior and memory effects are present also in the equilibrium (i.e. elastic) case. In this section we will review a procedure introduced by Lizana et al. [24] allowing to write down the equation of motion for the tracer of the elastic single file, where these memory effects are explicitly derived.

Let us start by considering with an overdamped Langevin equation for an harmonic chain:

$$\xi \frac{dx_n(t)}{dt} = \kappa \left[x_{n+1}(t) + x_{n-1}(t) - 2x_n(t) \right] + \eta_n(t). \tag{5.39}$$

Or, in the continuum limit:

$$\xi \frac{\partial x(n,t)}{\partial t} = \kappa \frac{\partial^2 x(n,t)}{\partial n^2} + \eta(n,t), \tag{5.40}$$

where, as usual, $\eta(n, t)$ is the Gaussian noise term of the n-th particle. Thanks to linearity, it is possible to easily pass to the Fourier and Laplace space, with the transformation:

$$\hat{x}(q,s) = \int_{-\infty}^{\infty} dn \int_0^{\infty} dt\, e^{-iqn-st} x(n,t). \tag{5.41}$$

Equation (5.40) gives

$$\hat{x}(q,s) = \frac{\hat{\eta}(q,s) + \xi \hat{x}(q,t=0)}{\xi s + \kappa q^2}. \tag{5.42}$$

Our purpose it to derive the equation of motion for a tracer particle. Let us consider, for instance, the particle $n = 0$. Subtracting $2\pi\delta(q)x(0, t = 0)/s$ from both sides of (5.42), rearranging, and taking the inverse Fourier transform at $n = 0$ one has:

$$\gamma(s)[s x_0(s) - x_0(t=0)] = \eta^{\text{eff}}(s), \tag{5.43}$$

where we have defined $x_0(s) \equiv x(0, s)$, in the Laplace space, and $x_0(t) \equiv x(0, t)$. Moreover, $\gamma(s) = \sqrt{4\xi\kappa/s}$ is a friction kernel, and the effective noise is defined as

$$\eta^{\text{eff}}(s) = \int_{-\infty}^{\infty} dn\, \exp(-\sqrt{\xi s/\kappa}|n|)[\eta(n,s) + \xi[x(n,t=0) - x_0(t=0)]]. \tag{5.44}$$

Note that the effective noise acting on the particle also includes the initial positions of the other particles.

The inverse Laplace transform of (5.43) yields

$$\sqrt{\frac{2\xi\kappa}{\pi}} \int_0^t \frac{dt'}{|t-t'|^{1/2}} \dot{x}_0(t') = \eta^{\text{eff}}(t), \tag{5.45}$$

Equation (5.45) is also known with the term “Fractional Langevin Equation” [21] because of the presence, at the left-hand term, of the *Caputo* derivative of order 1/2

$$\frac{d^{\alpha} f(t)}{dt^{\alpha}} = \frac{1}{\Gamma(1-\alpha)} \int_0^t \frac{dt'}{|t-t'|^{\alpha}} \frac{df(t')}{dt'}. \tag{5.46}$$

Note that Eq. (5.45) goes in the same direction of the ones derived in Chap. 4 for higher dimensions, namely the action of the surrounding particles changes both the friction and the noise terms. Remarkably, in this one dimensional case, memory effects does emerge also in the dilute regime, which are at the origin of the anomalous behavior of the mean square displacement.

In this example we have studied the harmonic case. Actually, for long time properties, the non-harmonic interaction are well described by an "harmonization procedure" [21]. Let us give an idea of this procedure: consider two particles in a unidimensional system with a large number N of particles in between, interacting via a non harmonic potential. If the distance between the two particles is L,one can introduce the free energy $F(L, N)$. Since N is large, fluctuations are supposed to be small, hence one can expand up to the second order in L:

$$\mathcal{F}(L, N) \sim \mathcal{F}(L_{\text{eq}}, N) + k_N (L - L_{\text{eq}})^2/2. \tag{5.47}$$

The harmonization procedure consists in replacing the non harmonic system with N harmonic springs in such a way that the free energy of the two systems, up to the second order in L, is the same. Therefore, if κ is the constant of the harmonic force of the N springs, from (5.47) one has:

$$\kappa = N k_N = N \frac{\partial^2 F}{\partial^2 L^2}. \tag{5.48}$$

In the case of hard-rods, for instance, using the van der Waals equation of state one has that

$$\kappa = \rho^2 k_B T (1 - \rho b)^{-2}. \tag{5.49}$$

In this way, by simply fixing the value of κ, The model (5.40) can be used to study the long time behavior of an hard rod system.

It must be noticed that, if one assumes thermal initial conditions for all the particles, then, also if geometrical constraints are present and the dynamics is anomalous, the system is in equilibrium. This aspect has important consequences:

- The mean square displacement can be calculated, yielding to $\langle \delta x^2(t) \rangle = \rho(1 - \sigma\rho)^{-1} \sqrt{4Dt/\pi}$, which is the same expression found also with other methods [1, 20, 30].
- The effective noise satisfies the fluctuation-dissipation relation

$$\left\langle \eta^{\text{eff}}(t) \eta^{\text{eff}}(t') \right\rangle = k_B T \gamma(|t - t'|). \tag{5.50}$$

- For a constant force acting on the tracer one can deduce the Einstein Relation

$$\langle x(t)\rangle_F = F\langle\delta x^2(t)\rangle_F/(2k_BT). \tag{5.51}$$

One may ask how it is possible that, also with this strong harmonic approximation it has been possible to obtain the exact prediction of some remarkable properties like the mean square displacement. Actually, the subdiffusion plays here a central role, indeed a difference in time scales is present: The time τ_{tr} which takes the tracer particle to cross a region of size L scales like L^4, on the contrary the time that the same region takes to reach equilibrium is proportional to L^2 which, for large L, is sensibly lower. In other words, the time scale separation induced by subdiffusion allows a sort of local equilibrium and makes the harmonization procedure possible.

5.3.2 The Response Relation

We are now going to study the linear response of the single file system. The dynamics of this system is highly non-trivial because of the long time tails in the autocorrelation of velocity but the steady state distribution function is known to be a Gaussian in the elastic case. Therefore, a study of the response must start from the generalized formula (2.18), that we recall here for clarity:

$$R_{i,j}(t) = \frac{\overline{\delta X_i(t)}}{\delta X_j(0)} = -\left\langle X_i(t)\left.\frac{\partial \ln \rho(\mathbf{X})}{\partial X_j}\right|_{t=0}\right\rangle, \tag{5.52}$$

where $\mathbf{X} = \{x_i, v_i\}$ is the collection of all the position and velocities of the particles. As already noticed above, in the elastic case,

$$\rho(\mathbf{X}) = \rho_x(\{x\})\prod_i \rho_v(v_i), \tag{5.53}$$

where ρ_v is the single-particle Maxwell-Boltzmann distribution. In summary, thanking to the phase space factorization and to the Gaussian distribution of velocities, one has that

$$\frac{\overline{\delta v_i(t)}}{\delta v_j(0)} = \beta\langle v_i(t)v_j(0)\rangle, \tag{5.54}$$

which is nothing but the Einstein relation in the differential form. As a consequence of the Gaussian nature of the problem, applying a perturbation as a small force in (5.35), one finds that the Einstein relation is always fulfilled. This result is not surprising in the case $L \to +\infty$ [26, 31, 38], since it is shown also in the analytic calculation of the previous section, but it holds also for finite N and L (see Fig. 5.7). The analogy with the comb model is now explicit: also in this case the fluctuation dissipation relation does not change in presence of the dynamical crossover of the mean square displacement.

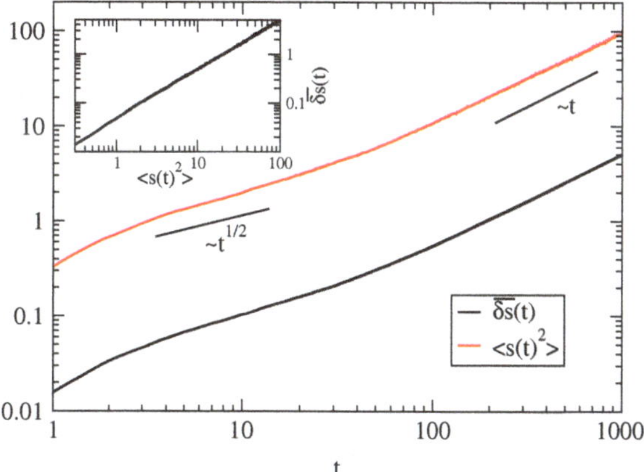

Fig. 5.7 $\langle s^2(t) \rangle$ and the response function $\overline{\delta s}(t)$ for the single-file model with parameters: $N = 10$, $L = 10$, $\sigma = 0.1$, $m = 1$, $\gamma = 2$, $T = 1$ and perturbation $F = 0.1$. In the *inset* the parametric plot $\overline{\delta s}(t)$ versus $\langle s^2(t) \rangle$ is shown. Also in this case, in analogy with the comb case, the generalized response relation is not sensible to a dynamical crossover experienced by a tracer particle

The main difference with the comb lattice resides in the Gaussian distribution of velocity. A straightforward consequence is evident if one consider a driven case, by adding a constant force F to the equations of motion. In this case, thanks to gaussianity, the naive correction (5.34) is largely verified in numerical simulations, also for finite N and L.

5.3.2.1 The Inelastic Case

Strong violations of the Einstein relation are observed in dense cases, when the collisions between the rods are inelastic: in this case the system is out of equilibrium and a net energy flux goes from the thermostat to the system and then dissipated in collisions [38].

We will not repeat here the analysis of the dilute case, since it is close to the ones discussed in Sect. 2.3.2. We only recall that, in a dilute regime the phase space factorization is still a good approximation, and that from (5.52) follows:

$$\frac{\overline{\delta v_i(t)}}{\delta v_j(0)} = -\left\langle v_i(t) \frac{\partial \ln \rho_v(v)}{\partial v_j} \bigg|_{t=0} \right\rangle. \tag{5.55}$$

In this case, $\rho_v(v)$ is a non Gaussian distribution of velocity, as described in Sect. 2.3.1 and therefore one could expect some deviations from the Einstein relation, due to higher order correlations. Actually, violations of Gaussianity have been shown in [31] to be not relevant for the fluctuation dissipation relation, because autocorrelations at different orders are almost proportional, i.e. $\langle v(0)v(t) \rangle / \langle v^2 \rangle \approx \langle v(0)^2 v(t) \rangle / \langle |v|^3 \rangle \approx$

$\langle v(0)^3 v(t)\rangle / \langle v^4 \rangle$ etc. This is confirmed by Direct Monte Carlo simulations, where an almost perfect factorization of the degrees of freedom in the phase-space probability distribution is satisfied: in such simulations, even with a stronger departure from Gaussianity, the Einstein relation always holds.

On the contrary, in a denser or in a strong dissipation regime, the phase space factorization fails, namely

$$\rho(\{x_i\}, \{v_i\}) \not\propto \prod_{i=1}^{N} \rho_x(\{x_i\}) p_v(v_i). \tag{5.56}$$

The main difference compared to the systems in higher dimension is that, due the non-overlapping property of the single file model, each particle has his own neighbors, and this simplify the possibility of characterizing the breakdown of phase-space factorization. A simple one is displayed in the inset of Fig. 5.8:

$$C_{v^2,v^2} = \frac{\langle \delta v_i^2 \delta v_{i+1}^2 \rangle}{\langle \delta v_i^4 \rangle}, \tag{5.57}$$

where $\delta v_i^2 = v_i^2 - T_g$. When $C_{v^2,v^2} > 0$, the squared velocities of two adjacent particles are correlated. It is evident that this correlation increases when α is decreased. The same is observed tuning the other parameters, such as decreasing r or increasing ϕ.

A more meaningful measure is given by the correlation

$$\mathcal{S}_{i,j} \equiv \langle v_i v_j \rangle. \tag{5.58}$$

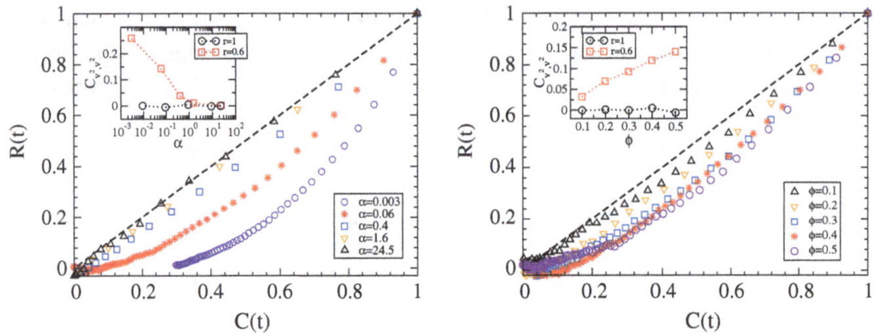

Fig. 5.8 Parametric plot of response $R(t)$ versus normalized autocorrelation $C(t)$. The *dashed line* is the Einstein relation $R \equiv C$. All data are obtained with restitution coefficient $r = 0.6$. On the *left* the packing fraction is constant $\phi = 0.1$ and τ_b is changed, resulting in different values of τ_c. The ratio $\alpha = \tau_c/\tau_b$ is given for simplicity. On the *right* $\tau_b = 1$ is kept constant, while ϕ is changed. In the *insets* the correlator C_{v^2,v^2}, discussed in the text, is displayed as a function of the varying parameter, for elastic and inelastic systems

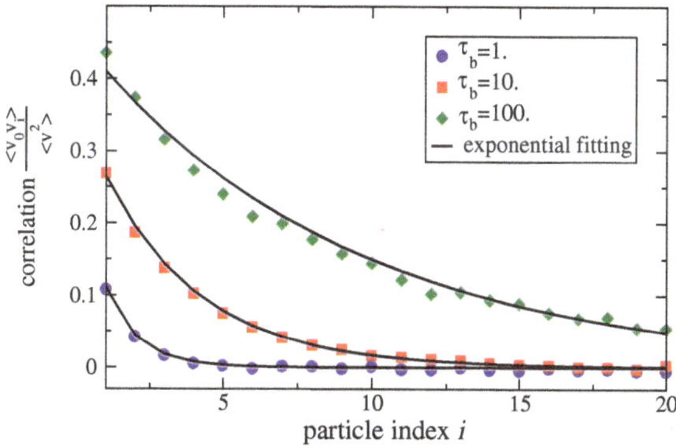

Fig. 5.9 Exponential dependence of $f(n)$ for different conditions of the value of $\tau_b = 1, 10, 100$. The data are normalized respect to $T_g \equiv \langle v^2 \rangle$. The *solid curves* are the best exponential fits. Values of the other parameters: $r = 0.6$, $T_b = 1$ and $\phi = 0.1$. The stronger is dissipation (i.e. the higher is τ_b) and the higher is the correlation among the particles

Clearly, since the particle are on a ring, the matrix $S_{i,j}$ has strong symmetries and one can write $S_{i,j} \equiv f(|i-j|)$, namely it is only a function of the "distance" between the particles. As shown in Fig. 5.9, from numerical data appear clear that $f(n = |i-j|) \simeq e^{-\frac{n}{\xi}}$, where ξ grows when dissipation increases. This result remembers similar correlation length introduced recently in higher dimensions at an hydrodynamic level [16, 17].

As a consequence, from Eq. (5.52) follows that the autocorrelation of velocity is not sufficient to have a correct prediction of the response, and this is translated in a *failure* of the equilibrium fluctuaction dissipation relation.

Given this considerations, the most natural way to proceed in the response analysis is to give some *ansatz* on the phase space distribution. It appear reasonable to assume a weak correlation between position and velocities and then, by applying (5.52) one has

$$\frac{\delta v_i(t)}{\delta v_j(0)} = -\left\langle v_i(t) \frac{\partial \ln \rho_v(\{v\})}{\partial v_j}\bigg|_{t=0} \right\rangle, \tag{5.59}$$

where $\rho_v(\{v\})$ is the velocity distribution of all the particles. Since non-Gaussianity does not play a key role, we can start with the approximation

$$\ln \rho_v(\{v\}) = -\frac{1}{2} \sum_{i,j} v_i \Sigma_{i,j} v_j, \tag{5.60}$$

where the matrix Σ is the inverse of the correlation matrix S defined in (5.58).

In order to fix ideas, let us consider to make a perturbation on the particle $i = 1$ and let us take, as first correction, the first neighbors (particles with index 0 and 2). The $S_{i,j}$ matrix reads

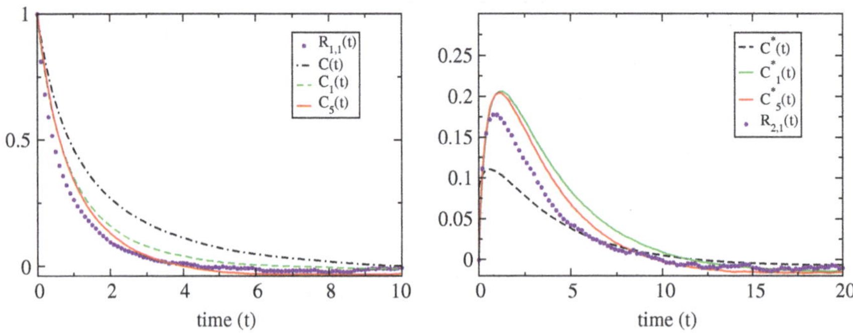

Fig. 5.10 *Left* Comparison between the response $\overline{\frac{\delta v_1(t)}{\delta v_1(0)}}$ and different correlations. *Right* Cross-response $\overline{\frac{\delta v_2(t)}{\delta v_1(0)}}$ veses Correlations. For sake of compactness, we have used with $C(t)$ and $C^*(t)$ the correlations predicted by the Einstein relation, with $C_2(t)$ and $C_2^*(t)$ the corrections expressed in (5.62), and with $C_5(t)$ and $C_5^*(t)$ the relative generalizations to five neighbors

$$S = \begin{pmatrix} \langle v^2 \rangle & \langle v_1 v_0 \rangle & \langle v_2 v_0 \rangle \\ \langle v_1 v_0 \rangle & \langle v^2 \rangle & \langle v_1 v_0 \rangle \\ \langle v_2 v_0 \rangle & \langle v_1 v_0 \rangle & \langle v^2 \rangle \end{pmatrix}, \tag{5.61}$$

where we have exploited the symmetries of the matrix S. By a straightforward application of formula (5.52)

$$\begin{aligned} \frac{\delta v_1(t)}{\delta v_1(0)} &= \Sigma_{1,1}\langle v_1(t)v_1(0)\rangle + 2\Sigma_{1,2}\langle v_1(t)v_0(0)\rangle \\ \frac{\delta v_2(t)}{\delta v_1(0)} &= \Sigma_{1,1}\langle v_2(t)v_1(0)\rangle + \Sigma_{1,2}\langle v_1(t)v_1(0)\rangle + \Sigma_{1,3}\langle v_1(t)v_0(0)\rangle. \end{aligned} \tag{5.62}$$

In a similar way one can introduce corrections beyond the first neighbors, obtaining a better prediction of the response, as shown in Fig. 5.10.

5.4 Ratchet Effects in Disordered Systems

As already discussed in Sect. 2.2.4, in an irreversible environment, thermal fluctuations can be rectified in order to produce a directed current. After a few fundamental examples of historical and conceptual value, in the last twenty years a huge amount of devices and models, usually known as Brownian ratchets or motors, have been proposed [2]. The purpose of these models is often practical, e.g. the extraction of energy from a highly fluctuating environment, such as a living cell [34]. But Brownian ratchets are also valid probes for the non-equilibrium properties of the fluctuating medium, the value of the current being sensitive to the interplay of different time-scales as well as different temperatures at work. In the following sections we will present a work inspired from the latter scenarios with a continuous flow of energy in a non-thermalized medium. The system under investigation is a fragile glass former and it has the peculiarity of being subdiffusive and non-stationary (see Sect. 2.2.1.4).

5.4.1 Description of the Model

The numerical experiments here proposed involves the 3D soft-spheres model, which is known to be a fragile glass-former [13, 29, 35]. We study a binary mixture (50:50) in a volume $V$of N particles with radii ratio, $\sigma_1/\sigma_0 = 1.2$. Particles interact via the soft potential $U(r) = [(\sigma_i + \sigma_j)/r]^{12}$ and the dynamics is evolved via a local Monte Carlo algorithm. Time is measured in Monte Carlo steps.[3]

The thermodynamic properties of soft-spheres are controlled by a single parameter $\Gamma = \rho T^{-1/4}$, which combines the temperature T and the density ρ of the system, with $\rho = N/V\sigma_0^3$ and σ_0 the radius of the effective one-component fluid. The model has a dynamical crossover at a mode-coupling temperature T_{MC} corresponding to the effective coupling $\Gamma_{MC} = 1.45$ [35].

In order to explore a ratchet effect, an intruder is introduced, with asymmetric interaction with all the other particles of the system. More specifically, denoting with x_i, $i > 0$ the x coordinate of the ith particle, and with x_0 the abscissa of the intruder, we choose

$$U(\mathbf{r}_0, \mathbf{r}_i) = \begin{cases} U(|\mathbf{r}_0 - \mathbf{r}_i|) & \text{if } \mathrm{x_i} < \mathrm{x_0}, \\ \varepsilon U(|\mathbf{r}_0 - \mathbf{r}_i|) & \text{otherwise}, \end{cases} \tag{5.63}$$

with $\varepsilon = 0.02$. The spatial symmetry along the x-axis is therefore broken, fulfilling, as we will see, one of the requirements to get a ratchet device.

The initial conditions are chosen from equilibrated configurations, embedding the asymmetric particle, at a high temperature $T_{liq} \gg T_{MC}$, where the system has simple liquid behavior with fast exponential relaxation. The dynamics of the asymmetric intruder is then studied both along equilibrium trajectories and after quenches to different temperatures $T < T_{MC}$.

Let us consider first the effect of the asymmetric interaction on the arrangement of particles around the intruder. We study the pair distribution function of neighbors to the right and to the left of the intruder,

$$g_L^R(r|\delta x \gtrless 0) = \sum_{j|\delta x_j \gtrless 0} \delta(r_{j0} - r), \tag{5.64}$$

with r_{j0} the distance between the intruder and the jth particle, in equilibrium or at a fixed elapsed time after the quench (see Fig. 5.11, left panels). Right neighbors stay closer to the intruder due to the reduced repulsion and this asymmetric clustering of neighbors is slightly enhanced during the aging regime.

[3] One step corresponds to N attempted Monte Carlo moves.

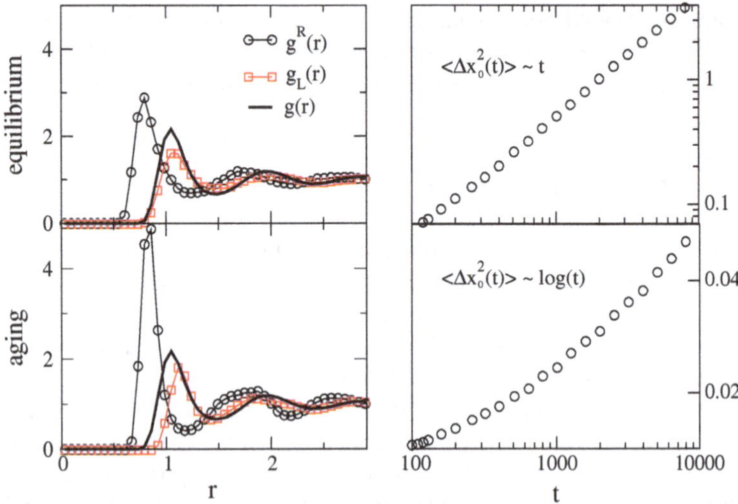

Fig. 5.11 *Left panels* Pair distribution function centered on the intruder, $g_L^R(r|\delta x \gtrless 0)$, of *left* (*red squares*), and *right* (*black circles*) neighbors, at equilibrium (*top*) and at a fixed elapsed time after quench (*bottom*). *Black line*, $g(r)$ for symmetric interactions. *Right panels* $\langle \Delta x_0^2(t) \rangle$ at equilibrium (*top*) and after the quench (*bottom*). The averages denoted by $\langle \ldots \rangle$ are realized considering 5000 initial configurations with a single intruder, equilibrated at $T_{liq} = 4.42\, T_{MC}$

5.4.1.1 The Ratchet Effect of the Asymmetric Intruder

Contrary to what observed in static measurements, dynamical measurements show important differences between the fluid and the aging regime. Let us consider the mean square displacement around the average position at time t of the intruder particle (Fig. 5.11, right): it is diffusive at high temperatures, while it is logarithmic after a quench, namely

$$\langle \Delta x_0^2(t) \rangle - \langle \Delta x_0(t) \rangle^2 \sim \log t. \tag{5.65}$$

This behavior, typical of activated dynamics in a rough potential [7], is also observed for host particles and it is a reflection of the critical slowing down of the dynamics after a quench well below the mode coupling temperature.

The behavior of the average displacement of the intruder is more striking, and it is possible to distinguish three cases:

- At equilibrium (Fig. 5.12, black circles) there is no net displacement on the x axes. In this case spatial symmetry is broken whereas time reversal symmetry is preserved.
- The same happens to $\langle \Delta y_0(t) \rangle$ (green diamonds) during aging: in this case only macroscopic time reversal symmetry is broken.

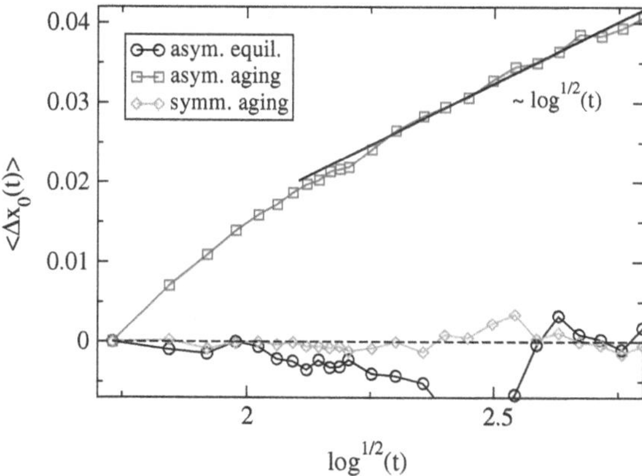

Fig. 5.12 Average intruder displacement: $\langle \Delta x_0(t) \rangle$ for equilibrium trajectories (*black circles*), and $\langle \Delta x_0(t) \rangle$ and $\langle \Delta y_0(t) \rangle$ after a quench far below T_{MC} ($T = 0.42T_{MC}$) (*red squares* and *green diamonds* respectively). The displacement $\langle \Delta x_0(t) \rangle$ is measured in units of the average inter-particle distance, as obtained from the position of the first peak in the pair distribution function $g(r)$ (see also Fig. 5.11)

- For $\langle \Delta x_0(t) \rangle$ after the quench (red squares), parity and time-reversal symmetry are both violated, and in this case a net average drift is found (Fig. 5.12), linear on a logarithmic timescale, $\langle \Delta x_0(t) \rangle \sim \tau$ with $\tau = \log^{1/2} t$.

This first exploration of ratcheting effects in glassy models opens two important issues, that deserves a deeper investigation. From one side, the role of activated events seems to be relevant and the main cause of the emerging logarithmic timescale of the drift. On the other side, ratchet is a pure non-equilibrium phenomena, and one expects that the dependence of the drift intensity on the parameters could give some information on the presence of a heat flowing into the system.

5.4.2 The Role of Activated Processes: The Asymmetric Sinai Model

In order to evaluate the generality of this glassy ratchet, it is important to reproduce it in a more controlled setup. Sinai model is a good candidate for this purpose: it is one of the simplest describing the diffusion of a single particle through a random correlated potential [8]. Its long-time dynamics is ruled by activated events and is characterized by a logarithmic time-scale.

In the original Sinai model the random potential is built from a random-walk of the force on a $1d$ lattice. The force F_i at each lattice site i is an independent identically distributed random variable extracted from a zero mean *symmetric* distribution $p(F)$, in this way the average force experienced by a particle along a trajectory is zero. The potential on a lattice site n is given by

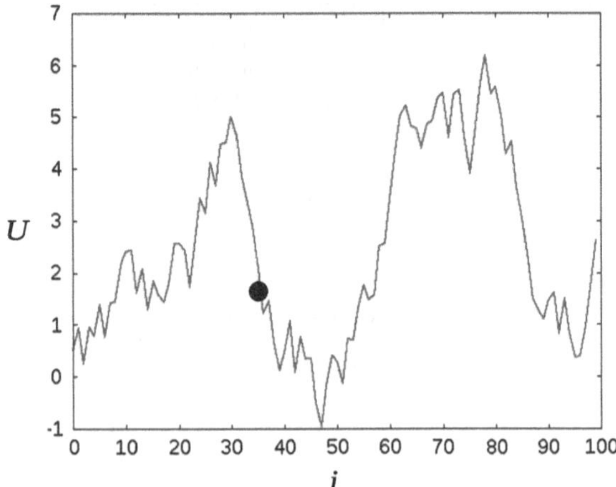

Fig. 5.13 Schematic representation of a random walk in a random potential

$$U(n) = \sum_{i=1}^{n} F_i. \tag{5.66}$$

If $\Delta l(t) = l(t) - l(0)$ is the displacement of the particle at time t, the long-time scaling of the mean square displacement is

$$\langle \Delta l^2(t) \rangle \sim \log^4(t). \tag{5.67}$$

The last equation can be obtained exploiting the simple following argument: the potential excursion between two sites grows like $\langle |U(i) - U(j)| \rangle \sim |i - j|^{1/2}$, such a relation, together with the expression of the typical time needed to jump a barrier, $\tau \sim \exp(\beta \Delta U)$, yields to the result (5.67). A mathematically rigorous demonstration of Eq. (5.67) has been first proposed in [36].

Logarithmic time scale for the growth of domain size is quite ubiquitous in the low temperature regime of glassy systems, where activated processes dominate [14].

In order to mimics the "equilibrium" dynamics at temperature $T = \frac{1}{\beta}$, in this model it is sufficient to average over initial conditions distributed as

$$P_0(i) \sim \exp[-\beta U(i)]. \tag{5.68}$$

On the contrary, the time-reversal symmetry breaking can be induced with the choice of a different initial condition. An initial infinite temperature $T_{liq} = \infty$ is obtained by extracting the starting point of the particle from a flat distribution. The quench to a glassy-like phase is then obtained evolving the system at a temperature $T \ll \sqrt{L}$, where L is the linear size of the system. We have studied a "spatially asymmetric" version of the Sinai model, where the symmetric force distribution $p(F)$ is replaced with an asymmetric one $\tilde{p}(F)$, in analogy with the asymmetric potential of the glassy ratchet, in order break the spatial symmetry. The asymmetric random force is obtained according to the following procedure:

- at each site the sign of a random variable f_i is chosen with probability $1/2$,
- its modulus is chosen from an exponential probability distribution $\frac{1}{k}e^{-|f_i|/k}$ where k equal to λ_+ or λ_-, depending on the sign extracted.
- the potential is then built as $U(n) = \sum_{i=1}^{n} F_i$, with $F_i = f_i - (\lambda_+ - \lambda_-)/2$. This amounts to a shift of the whole distribution such that $\langle F \rangle = 0$.

The results of the simulations of this model are shown in Fig. 5.14: they clearly show a behavior in striking similarity with those of Fig. 5.12 for the glass-former model: a drift is observed only when both time reversal and spatial symmetry are broken. Indeed, it is sufficient that one of the two symmetries is restored to have a zero drift: in this case, even with the asymmetric distribution of forces described above, the average drift is zero. We conclude by noting that the large time behavior of the drift is compatible with a squared logarithm

$$\langle \Delta l(t) \rangle \sim \log^2(t) \sim \sqrt{\langle \Delta l^2(t) \rangle}, \tag{5.69}$$

similarly to what observed previously for the glassy ratchet. The Sinai model is made of few essential ingredients, so that the finding of a net displacement suggests us that the glassy ratchet phenomena studied here is not model dependent. The relevance of activated events for non equilibrium transport properties clearly appears from the logarithmic drift observed in both the Sinai and soft spheres model.

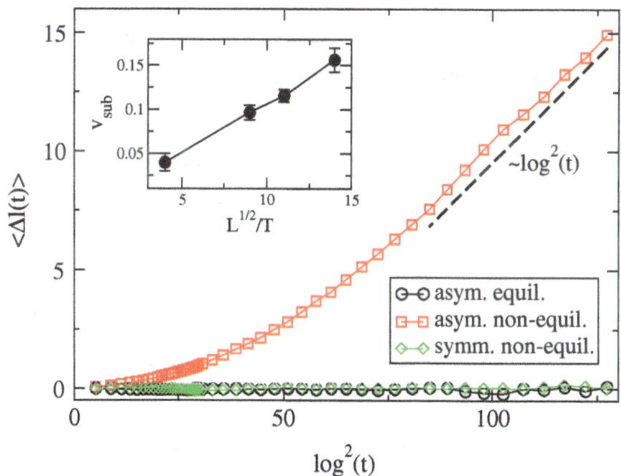

Fig. 5.14 *Main* Average displacement in Sinai model. *Black circles* diffusion in asymmetric potential at equilibrium *green diamonds* diffusion during aging in symmetric potential (quench from $T = \infty$) *red squares* diffusion during aging in asymmetric potential. *Inset* the sublinear drift velocity v_{sub} grows with $\sqrt{L}/T$ (see Sect. 5.4.3). In the present simulation $\lambda_+ = 1$ and $\lambda_- = 0.2$ have been used

5.4.3 The Two Temperature Scenario

Typical examples of ratchets in ideally statistically stationary configurations are obtained by coupling the system with two or more reservoirs, as described in Sect. 2.2.4. It is the existence of different temperatures within the same system which allows the production of work without violations of the second principle of thermodynamics. But what is the second temperature in the glassy ratchet? According to the well-established description of the aging regime of glasses [9], it is the effective temperature, defined as the violation factor of the fluctuation-dissipation theorem.

The quench of a fragile glass, for instance our soft spheres model, below its mode-coupling temperature produces aging and violations. As shown in the introductory Sect. 2.2.1.4 it is quite common to describe the system in terms of a generalized response relation of the kind

$$T\chi(t, t_w) = X(t, t_w)[C(t, t) - C(t, t_w)], \tag{5.70}$$

with $\chi(t, t_w)$ the integrated response and $C(t, t_w)$ the correlation, yielding the definition of the effective temperature $T_{eff}(t, t_w) = T/X(t, t_w)$. The last is usually higher than the bath temperature $T_{eff}(t, t_w) > T$ and is understood as the temperature of slow, still not equilibrated, modes. Clearly, the ratio T_{eff}/T may be regarded as the parameter which tunes non equilibrium effects and we study here how the glassy ratchet drift depends on it. We obtain T_{eff} from the parametric plot of $C(t, t_w)$, taken as the self-intermediate scattering function, versus the integrated response $T\chi(t, t_w)$, as shown in Fig. 5.15, measured according to the field-free method of [5].

Once detected the possible second temperature into the system, we need to extract from the curve $\langle \Delta x_0(t) \rangle$ a synthetic observable. The finding of a drift on a logarithmic timescale, suggests to define an average "sub-velocity" as:

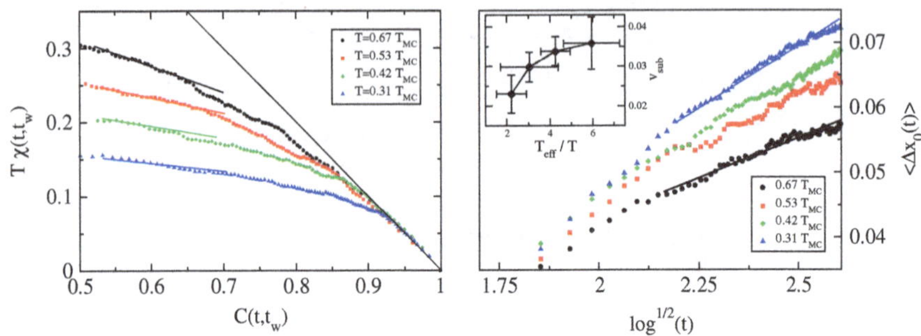

Fig. 5.15 *Left panel* Parametric plot $T\chi(t, t_w)$ versus $C(t, t_w)$. *Right panel* asymmetric intruder drift for different quench temperatures; linear fits yield sublinear velocities. *Inset* v_{sub} versus T_{eff}/T. In both *panels* colors of data correspond to different effective temperatures: T_{eff} = $1.497T_{MC}$ (*black*), $1.613T_{MC}$ (*red*), $1.792T_{MC}$ (*green*) and $1.819T_{MC}$ (*blue*)

$$v_{sub}(t, t_w) = \frac{\langle \Delta x_0(t) \rangle - \langle \Delta x_0(t_w) \rangle}{\delta \tau} = \frac{\langle x_0(t) \rangle - \langle x_0(t_w) \rangle}{\delta \tau}, \quad (5.71)$$

with $\delta\tau = \log^{1/2} t - \log^{1/2} t_w$. This average subvelocity depends in general on both the running time t and the waiting time t_w elapsed since the quench. More precisely, considering Fig. 5.12, by fixing t_w and t we choose the time lag where the slope of the curve $\langle \Delta x_0(t) \rangle$ is measured. Clearly, the "instantaneous" sub-velocity only depends on t_w and corresponds to a slowly decaying velocity[4]

$$v(t_w) = d\langle \Delta x_0(t_w) \rangle / dt_w \sim 1/t_w. \quad (5.72)$$

For large enough waiting times we can define the "order parameter" v_{sub}, namely we find $v_{sub}(t_w) \sim$ const, and then we probe its dependence upon the external parameters. Here we focus on the quench temperature T, drawing a connection between the drift of the glassy ratchet, which is a pure non-equilibrium effect, and the more customary equilibrium-like descriptions of aging media in terms of effective temperatures.

The inset of the right panel of Fig. 5.15 shows the behavior of v_{sub} *verses* T_{eff}/T, revealing that the subvelocity increases when T_{eff}/T is increased. Namely the intensity of the ratchet effect, traced in the measure of v_{sub}, grows as the distance from equilibrium is increased.

The out-of-equilibrium dynamics of the Sinai model closely reproduces these observations: in the inset of Fig. 5.14 shows the sub-velocity as a function of $\sqrt{L}/T$, with $T \ll L^{1/2}$ the quench temperature and L the size of the linear chain. The inset of Fig. 5.14 shows that v_{sub}, signaling that also for the Sinai model the larger the distance from equilibrium the larger the velocity of the drift.

References

1. Alexander, S., Pincus, P.: Diffusion of labeled particles on one-dimensional chains. Phys. Rev. B **18**, 2011 (1978)
2. Astumian, R.: Thermodynamics and kinetics of a brownian motor. Science **276**, 917 (1997)
3. Baiesi, M., Boksenbojm, E., Maes, C., Wynants, B.: Nonequilibrium linear response for markov dynamics, ii: Inertial dynamics. J. Stat. Phys. **139**, 492 (2010)
4. Baiesi, M., Maes, C., Wynants, B.: Nonequilibrium linear response for markov dynamics, i: jump processes and overdamped diffusions. J. Stat. Phys. **137**, 1094 (2009)
5. Berthier, L.: Efficient measurement of linear susceptibilities in molecular simulations: application to aging supercooled liquids. Phys. Rev. Lett. **98**, 220601 (2007)
6. Biferale, L., Crisanti, A., Vergassola, M., Vulpiani, A.: Eddy diffusivities in scalar transport. Phys. Fluids **7**, 2725 (1995)
7. Bouchaud, J.: Weak ergodicity breaking and aging in disordered systems. J. de Phys. **I**(2), 1705 (1992)
8. Bouchaud, J., Georges, A.: Anomalous diffusion in disordered media: statistical mechanisms, models and physical applications. Phys. Rep. **195**, 127 (1990)

[4] With logarithmic corrections.

9. Bouchaud, J.P., Cugliandolo, L., Kurchan, J., Mezard, M.: Mode-coupling approximations, glass theory and disordered systems. Physica A: Statistical Mechanics and its Applications, **226**(3), 243–273 (1996)
10. Burioni, R., Cassi, D., Giusiano, G., Regina, S.: Anomalous diffusion and hall effect on comb lattices. Phys. Rev. E **67**, 016116 (2003)
11. Castiglione, P., Mazzino, A., Muratore-Ginanneschi, P., Vulpiani, A.: On strong anomalous diffusion. Physica D **134**, 75 (1999)
12. Cecconi, F., Diotallevi, F., Marconi, U.M.B., Puglisi, A.: Fluid-like behavior of a one-dimensional granular gas. J. Chem. Phys. **120**, 35 (2004)
13. Coluzzi, B., Mézard, M., Parisi, G., Verrocchio, P.: Thermodynamics of binary mixture glasses. J. Chem. Phys. **111**, 9039 (1999)
14. Corberi, F., Cugliandolo, L., Yoshino, H.: Growing length scales in aging systems. Dynamical heterogeneities in glasses, colloids, and granular media, p. 370 (2011)
15. Fisher, M.: Shape of a self-avoiding walk or polymer chain. J. Chem. Phys. **44**, 616 (1966)
16. Gradenigo, G., Sarracino, A., Villamaina, D., Puglisi, A.: Growing non-equilibrium length in granular fluids: from experiment to fluctuating hydrodynamics. Europhys. Lett. **96**, 14004 (2011)
17. Gradenigo, G., Sarracino, A., Villamaina, D., Puglisi, A.: Non-equilibrium length in granular fluids: From experiment to fluctuating hydrodynamics. Europhys. Lett. **96**, 14004 (2011)
18. Gu, Q., Schiff, E.A., Grebner, S., Wang, F., Schwarz, R.: Non-gaussian transport measurements and the Einstein relation in amorphous silicon. Phys. Rev. Lett. **76**, 3196 (1996)
19. Jepps, O.G., Rondoni, L.: Thermodynamics and complexity of simple transport phenomena. J. Phys. A **39**, 1311 (2006)
20. Kärger, J.: Straightforward derivation of the long-time limit of the mean-square displacement in one-dimensional diffusion. Phys. Rev. A **45**, 4173 (1992)
21. Lee, M.H.: Ergodic theory, infinite products, and long time behavior in hermitian models. Phys. Rev. Lett. **87**, 250601 (2001)
22. Levitt, D.: Dynamics of a single-file pore: Non-fickian behavior. Phys. Rev. A **8**, 3050 (1973)
23. Lippiello, E., Corberi, F., Sarracino, A., Zannetti, M.: Nonlinear susceptibilities and the measurement of a cooperative length. Phys. Rev. B **77**, 212201 (2008)
24. Lizana, L., Ambjörnsson, T., Taloni, A., Barkai, E., Lomholt, M.: Foundation of fractional langevin equation: harmonization of a many-body problem. Phys. Rev. E **81**, 051118 (2010)
25. Mancinelli, R., Vergni, D., Vulpiani, A.: Front propagation in reactive systems with anomalous diffusion. Physica D **185**, 175 (2003)
26. Marconi, U.M.B., Puglisi, A., Rondoni, L., Vulpiani, A.: Fluctuation-dissipation: response theory in statistical physics. Phys. Rep. **461**, 111 (2008)
27. Metzler, R., Barkai, E., Klafter, J.: Anomalous diffusion and relaxation close to thermal equilibrium: a fractional Fokker-Planck equation approach. Phys. Rev. Lett **82**, 3563 (1999)
28. Metzler, R., Klafter, J.: The random walk's guide to anomalous diffusion: a fractional dynamics approach. Phys. Rep. **339**, 1 (2000)
29. Parisi, G.: Off-equilibrium fluctuation-dissipation relation in fragile glasses. Phys. Rev. Lett. **79**, 3660 (1997)
30. Percus, J.: Anomalous self-diffusion for one-dimensional hard cores. Phys. Rev. A **9**, 557 (1974)
31. Puglisi, A., Baldassarri, A., Vulpiani, A.: Violations of the Einstein relation in granular fluids: the role of correlations. J. Stat. Mech. P08016 (2007)
32. Redner, S.: A guide to first-passage processes. Cambridge University Press, Cambridge (2001)
33. Richardson, L.F.: Atmospheric diffusion shown on a distance-neighbour graph. Royal Soci. Lond. Proc. Ser. A **110**, 709 (1926)
34. Rousselet, J., Salome, L., Ajdari, A., Prost, J.: Directional motion of brownian particles induced by a periodic asymmetric potential. Nature **370**, 446 (1994)
35. Roux, J., Barrat, J., Hansen, J., et al.: Dynamical diagnostics for the glass transition in soft-sphere alloys. J. Phys.: Condens. Matter **1**, 7171 (1989)

36. Sinai, Y.: Limiting behavior of a one-dimensional random walk in a random medium. Theory Prob. Appl. **27**, 256 (1983)
37. Trefan, G., Floriani, E., West, B.J., Grigolini, P.: Dynamical approach to anomalous diffusion: response of Levy processes to a perturbation. Phys. Rev. E **50**, 2564 (1994)
38. Villamaina, D., Puglisi, A., Vulpiani, A.: The fluctuation-dissipation relation in sub-diffusive systems: the case of granular single-file diffusion. J. Stat. Mech. L10001 (2008)
39. Weiss, G., Havlin, S., Bunde, A.: On the survival probability of a random walk in a finite lattice with a single trap. J. Stat. Phys. **40**, 191 (1985)

Chapter 6
Conclusions and Perspectives

Abstract This chapter is devoted to collect the main results presented in the rest of the work and to put them in a more unified point of view. The main role of cross correlations in non-equilibrium systems is pointed out and some attention is posed on possible perspectives and future works, as in the case of hydrodynamical fluctuations.

Let us summarize, in a schematic fashion, the results achieved in this work:

- **Projection operations and entropy production**: The effect of projections on entropy production is investigated. More specifically an equation with memory and colored noise is compared with the equivalent Markovian model. The latter leads to identify memory as a non-conservative force. These forces cease to contribute to entropy production only under the validity of the fluctuation dissipation relation of the second kind, which is equivalent to the detailed balance assumption. The projection from the Markovian to the non Markovian representation produces a loss of information detected from a decrease in mean entropy production. This change can be dramatic in the linear case, leading to the false conclusion that the system is in equilibrium.
- **Local velocity field in a granular gas**: We designed a first granular dynamical theory describing non-equilibrium correlations and responses for a massive tracer. In the dilute regime, under the molecular chaos assumption, the tracer dynamics is Markovian and stationary, and the equation satisfies detailed balance, also if inelasticity is present. On the contrary, in a denser regime the dynamics of the tracer is non Markovian and memory effects are present. The equation with memory is introduced and gives a significant insight into the mechanisms of recollision and dynamical memory and their relation with the breakdown of equilibrium properties. It is remarkable that velocity correlations between the intruder and the surrounding velocity field, in the inelastic case, are responsible for both the violations of the equilibrium fluctuation dissipation relation and the appearance of a non-zero entropy production.

D. Villamaina, *Transport Properties in Non-Equilibrium and Anomalous Systems*, Springer Theses, DOI: 10.1007/978-3-319-01772-3_6,

- **Einstein relations in subdiffusive models**: we have considered two systems with subdiffusive behavior, showing that the proportionality between response function and correlation breaks down when "non-equilibrium" conditions are introduced. In the case of a random walk on a comb model the response relation can be explicitly written, providing the out of equilibrium corrections to the Einstein relation. In the single file model an explicit formula is not available but we have shown that, by taking into account the coupling of the velocities of neighboring particles, it is possible to have a better prediction of the response.
- **The glassy ratchet as a non-equilibrium thermometer**: Through numerical simulations in different models and different choices of the quench temperature, always chosen in the deep slowly relaxing regime, the "glassy ratchet" phenomenon is investigated. The drift velocity slowly decays in time and can be appreciably different from zero. The overall intensity of the drift, measured in terms of a "subvelocity", is monotonically increasing with the distance from equilibrium, namely with the difference between the quench and effective temperatures. This observation supports the idea of regarding the ratchet drift as a "non-equilibrium thermometer": it can be used as a device capable to say how far a system is from equilibrium.

In summary, the general message that one learns is that correlations among different degrees of freedom can work as a channel of energy exchange, being responsible of the breaking of the equilibrium fluctuation dissipation relation and acting as a primary source of entropy production. In a non-equilibrium context, a too strong projection operation and the consequent reduction of information leads to a reduced description with, in some cases, a vanishing entropy production.

It must be noticed that, even if the scenario described appears very general, it has been tested in a quite controlled setup like the massive intruder in a granular gas model. The energy injection mechanism is homogeneous and the intruder follows a Langevin dynamics. One may wonder if this description is valid also for more complex situations. The Chap. 5 has been written as partial response to this general question. The subdiffusive models analyzed, even if with some differences, confirm the interpretations given for higher dimensions. A relevant ingredient in breaking the Einstein formula, for stationary regimes, is not the anomalous diffusion but the presence of currents driving the system out of equilibrium. The generalized response relations are a good tool to detect the main sources of non-equilibrium present in the system. In this direction, in the single file model a non-equilibrium correlation length can be defined and measured with simplicity, in analogy with the length coupled to a massive intruder in higher dimensions. However, this is only a partial result deserving a deeper investigation, since an equation for the tracer in the granular single file model is lacking and entropy production for this model is a challenging issue.

A very promising direction for future researches is the study of fluctuating hydrodynamics, already faced in some recent works [1]. The hydrodynamic equations are projections on slow modes and, as a consequence, they are associated to a loss of information. On the other hand, different models at a microscopic level can have the same hydrodynamic description, therefore the non-equilibrium currents "surviving" to a hydrodynamic projection are, in a sense, supposed to be more general with respect to the microscopic details of the model under investigation. Crucial, with respect to this "universality", is the role of microscopic details in the specific form of non-equilibrium fluctuations, which appear as noise in fluctuating hydrodynamics.

Reference

1. Gradenigo, G., Sarracino, A., Villamaina, D., Puglisi, A.: Growing non-equilibrium length in granular fluids: from experiment to fluctuating hydrodynamics. Europhys. Lett. **96**, 14004 (2011)

Appendix

1 Generalized Response Relation and Detailed Balance Condition

In this section we will derive the generalized response formula introduced in Sect. 2.2.1.2.

Let us consider a path ω for $s \in [0, t]$, and let us define the probability density

$$\mathcal{P}(\omega) \equiv \frac{\text{Prob}^h[\omega]}{\text{Prob}[\omega]} = e^{-\mathcal{A}}. \tag{1}$$

The action $\mathcal{A}$ can be decomposed in two parts, introducing the time-reversal $\mathcal{I}$ which changes ω into $(\mathcal{I}\omega)_s$ (for instance it changes the sign to the velocities)

$$\mathcal{A} = \frac{1}{2}(\mathcal{T}(\omega) - \mathcal{S}(\omega)), \tag{2}$$

where

$$\begin{aligned} \mathcal{T}(\omega) &= \mathcal{A}(\mathcal{I}\omega) + \mathcal{A}(\omega), \\ \mathcal{S}(\omega) &= \mathcal{A}(\mathcal{I}\omega) - \mathcal{A}(\omega). \end{aligned} \tag{3}$$

Then

$$\mathcal{P}(\omega) = e^{-\mathcal{T}(\omega)/2} e^{\mathcal{S}(\omega)/2}. \tag{4}$$

The anti-symmetric part $\mathcal{S}$ represents the excess in physical entropy due to perturbation, whereas the time symmetric term $\mathcal{T}$ is the excess in the time-integrated instantaneous of a quantity called dynamical activity [1, 2]. Notice that Eq. (3) can be also rewritten as

D. Villamaina, *Transport Properties in Non-Equilibrium and Anomalous Systems*, Springer Theses, DOI: 10.1007/978-3-319-01772-3,

$$\mathcal{S} = \log \left(\frac{\text{Prob}[\mathcal{I}\omega]}{\text{Prob}[\omega]} \Big/ \frac{\text{Prob}^h[\mathcal{I}\omega]}{\text{Prob}^h[\omega]} \right),$$

$$\mathcal{T} = \log \frac{\text{Prob}[\mathcal{I}\omega]\text{Prob}[\omega]}{\text{Prob}^h[\mathcal{I}\omega]\text{Prob}^h[\omega]}. \tag{5}$$

For a given observable Q, the response function can be then written

$$R_{QV}(t't') \equiv \frac{\delta\langle Q(t)\rangle_h}{\delta h(t')} = \frac{\delta\langle Q(t)e^{-\mathcal{T}/2+\mathcal{S}/2}\rangle}{\delta h(t')}$$

$$= \frac{1}{2}\left\langle Q(t)\, \frac{\delta\mathcal{S}}{\delta h(t')}\bigg|_{h=0}\right\rangle - \frac{1}{2}\left\langle Q(t)\, \frac{\delta\mathcal{T}}{\delta h(t')}\bigg|_{h=0}\right\rangle. \tag{6}$$

Let us consider a continuous-time homogeneous Markov process with transition rates $W(y|x)$ between states x and y and persistence probability of remaining in the state x for a time t, $P(x;t)$. As perturbed process, we consider the one evolving in presence of the external field $h(t)$ coupled to the observable $V(x)$. The persistence probabilities get changed into $P_h(x;t)$.

For the perturbed transition rates[1] $W_h(y|x)$ we assume the local detailed balance condition

$$W_h(y|x) = W(y|x)e^{\beta/2h(t)[V(y)-V(x)]}. \tag{7}$$

Then let us write explicitly the probability of a trajectory and of its time-reversal

$$\begin{aligned}
\text{Prob}[\omega] &= \rho(x_0)P(x_0;t_0)W(x_1|x_0)P(x_1;t_1)W(x_2|x_1)\ldots \\
&\quad \times W(x_n|x_{n-1})P(x_n;t_n) \\
\text{Prob}[\mathcal{I}\omega] &= \rho(x_n)P(x_n;t_n)W(x_{n-1}|x_n)P(x_{n-1};t_{n-1})W(x_{n-2}|x_{n-1})\ldots \\
&\quad \times W(x_0|x_1)P(x_0;t_0) \\
\text{Prob}^h[\omega] &= \rho(x_0)P(x_0;t_0;h(t_0))W(x_1|x_0;h(t_0))P(x_1;t_1;h(t_1)) \\
&\quad \times W_h(x_2|x_1)\ldots W(x_n|x_{n-1};h(t_{n-1}))P(x_n;t_n;h(t_n)) \\
\text{Prob}^h[\mathcal{I}\omega] &= \rho(x_n)P(x_n;t_n;h(t_n))W(x_{n-1}|x_n;h(t_n)) \\
&\quad \times P(x_{n-1};t_{n-1};h(t_{n-1}))\ldots W(x_0|x_1;h(t_1))P(x_0;t_0;h(t_0)),
\end{aligned} \tag{8}$$

where ρ is the distribution of the initial conditions. Now we can give an explicit meaning to the entropy and frenesy excesses of Eq. (5). Let us start by computing the entropic contribution. Enforcing the Local detailed balance condition (7) one obtains

[1] We have assumed the short hand notation $W_h(y|x) \equiv W(y|x;h(t))$ and $P_h(x;t) \equiv P(y|x;h(t))$, where it is not ambiguous.

$$\frac{\text{Prob}^h[\omega]}{\text{Prob}^h[\mathcal{I}\omega]} = \frac{\rho(x_0)}{\rho(x_n)} \exp\left\{\beta \int_0^t ds\, h(s)\dot{V}(s)\right\} \times \frac{W(x_1|x_0)W(x_2|x_1)\ldots W(x_n|x_{n-1})}{W(x_{n-1}|x_n)W(x_{n-2}|x_{n-1})\ldots W(x_0|x_1)} \tag{9}$$

and

$$\frac{\text{Prob}[\omega]}{\text{Prob}[\mathcal{I}\omega]} = \frac{\rho(x_0)}{\rho(x_n)} \frac{W(x_1|x_0)W(x_2|x_1)\ldots W(x_n|x_{n-1})}{W(x_{n-1}|x_n)W(x_{n-2}|x_{n-1})\ldots W(x_0|x_1)}. \tag{10}$$

Hence,

$$S = \log\left(\frac{\text{Prob}^h[\omega]}{\text{Prob}^h[\theta\omega]}\right) - \log\left(\frac{\text{Prob}[\omega]}{\text{Prob}[\theta\omega]}\right) = \beta \int_0^t ds\, h(s)\dot{V}(s). \tag{11}$$

The "frenetic" term can be obtained in the same way:

$$\begin{aligned} \mathcal{T} &= -\log \frac{\text{Prob}^h[\mathcal{I}\omega]\text{Prob}^h[\omega]}{\text{Prob}[\mathcal{I}\omega]\text{Prob}[\omega]} \\ &= -\log \frac{[P(x_0; t_0; h(t_0))\ldots P(x_n; t_n; h(t_n))]^2}{[\ldots P(x_n; t_n)]^2} \\ &= -2\log P(x_0; t_0; h(t_0))P(x_1; t_1; h(t_1))\ldots P(x_n; t_n; h(t_n)) \\ &\quad + 2\log P(x_0; t_0)P(x_1; t_1)\ldots P(x_n; t_n). \end{aligned} \tag{12}$$

Using the definition of persistence probability

$$P(x; t; h(t)) = e^{-\sum_{y\neq x}\int_0^t ds\ W(y|x;h(s))}, \tag{13}$$

together with the local detailed balance condition one obtains

$$\mathcal{T} = 2\int_0^t ds \left[\sum_{y\neq x} W(y|x)\left(e^{\beta/2h(s)[V(y)-V(x)]} - 1\right)\right], \tag{14}$$

and then

$$\left\langle Q(t)\, \frac{\delta\mathcal{T}}{\delta h(t')}\bigg|_{h=0}\right\rangle = \beta\langle Q(t)B(t')\rangle, \tag{15}$$

where

$$B(t) \equiv \sum_{y\neq x} W(y|x)[V(y) - V(x(t))]. \tag{16}$$

In the end, for the response function one finds

$$R_{QV}(t,t') = \frac{\beta}{2}[\langle Q(t)\dot{V}(t')\rangle - \langle Q(t)B(t')\rangle]. \tag{17}$$

2 Entropy Production for a System with Memory

Consider the following simple one-dimensional Langevin equation

$$m\ddot{x} = -\gamma\dot{x} + -h[x] + \eta, \tag{18}$$

where $\eta(t)$ is Gaussian noise of zero mean and correlation

$$\langle \eta(t)\,\eta(t')\rangle = \nu(t-t'), \tag{19}$$

with $\nu(t) = \nu(-t)$. The force term $h[x]$ contains a local in time part, denoted h_x, and a linear memory term,

$$h[x(t)] = h_x[x(t)] - \int_{-\infty}^{t} dt'\, g(t-t')\, x(t'). \tag{20}$$

Both $\nu(t)$ and $g(t)$ are left unspecified.

Since we are interested into the stationary regime, we let the initial time to $-\infty$, and the final one to $+\infty$. Under this assumption the probability of the a trajectory generated by the Langevin equation (18) is

$$\mathcal{P}\{x\} \propto \exp\left\{-\frac{1}{2}\int_{-\infty}^{+\infty} dt\, dt'\, X[x(t)]\nu^{-1}(t-t')X[x(t')]\right\}, \tag{21}$$

where $\nu^{-1}(t)$ is the inverse of $\nu(t)$ defined as

$$\int_{-\infty}^{+\infty} ds\, \nu(t-s)\,\nu^{-1}(s-t') = \int_{-\infty}^{+\infty} ds\, \nu^{-1}(t-s)\,\nu(s-t') = \delta(t-t'), \tag{22}$$

and

$$X[x(t)] = m\ddot{x}(t) + \gamma\dot{x}(t) + h[x(t)]. \tag{23}$$

By going in the Fourier space,

$$x(t) = \int_{-\infty}^{+\infty} \frac{d\omega}{2\pi}\, e^{-i\omega t}\, x(\omega) \quad\longleftrightarrow\quad x(\omega) = \int_{-\infty}^{+\infty} dt\, e^{i\omega t}\, x(t), \tag{24}$$

the probability (21) becomes

$$\mathcal{P}\{x\} \propto \exp\left\{-\frac{1}{2}\int_{-\infty}^{+\infty}\frac{d\omega}{2\pi}\,\widetilde{X}[x(\omega)]\nu(\omega)^{-1}\widetilde{X}[x(-\omega)]\right\}, \tag{25}$$

where $\nu^{-1}(\omega) = 1/\nu(\omega)$, with $\nu(-\omega) = \nu(\omega)$, and

$$\widetilde{X}[x(\omega)] = \omega^2 x(\omega) - i\omega x(\omega) + h(\omega) \tag{26}$$

with

$$h(\omega) = \int_{-\infty}^{+\infty} dt\, e^{i\omega t}\, h[x(t)]. \tag{27}$$

Consider now the reversed trajectory $x^{\mathrm{R}}(t) = x(-t)$. Its probability follows from (25) by noticing that $x^{\mathrm{R}}(\omega) = x(-\omega)$. To compute the ratio between the probability of a trajectory x and its reversed x^{R} we then have to separate the terms even and odd under the replacement $x(\omega) \to x(-\omega)$ into (25). To this end we have to look closer to $h(\omega)$.

From its definition we have

$$h(\omega) = h_x(\omega) - \int_{-\infty}^{+\infty} dt\, e^{i\omega t}\int_{-\infty}^{t} dt'\, g(t-t')\, x(t'). \tag{28}$$

Now

$$\begin{aligned}\int_{-\infty}^{t} dt'\, g(t-t')\, x(t') &= \int_{-\infty}^{+\infty}\frac{d\omega}{2\pi}\, x(\omega)\int_{-\infty}^{t} dt'\, e^{-i\omega t'}\, g(t-t')\\ &= \int_{-\infty}^{+\infty}\frac{d\omega}{2\pi}\, e^{-i\omega t}\, x(\omega)\int_{0}^{\infty} dt'\, e^{i\omega t'}\, g(t'),\end{aligned} \tag{29}$$

so that

$$h(\omega) = h_x(\omega) - g(\omega)\, x(\omega), \tag{30}$$

with

$$\begin{aligned} g(\omega) &= \int_0^{\infty} dt\, e^{i\omega t}\, g(t)\\ &= \int_0^{\infty} dt'\, \cos(\omega t)\, g(t) + i\int_0^{\infty} dt\, \sin(\omega t)\, g(t)\\ &= \phi(\omega) + i\omega\,\psi(\omega),\end{aligned} \tag{31}$$

where

$$\phi(\omega) = \int_0^{\infty} dt'\, \cos(\omega t)\, g(t) \tag{32}$$

$$\psi(\omega) = \int_0^\infty dt\, \frac{\sin(\omega t)}{\omega}\, g(t), \tag{33}$$

are real even functions of ω. Collecting all terms we have

$$h(\omega) = h_x(\omega) - \phi(\omega)\, x(\omega) - i\omega\, \psi(\omega)\, x(\omega) \tag{34}$$

and (25) takes the form

$$\begin{aligned}
\mathcal{P}\{x\} &\propto \exp\left\{-\frac{1}{2}\int_{-\infty}^{+\infty}\frac{d\omega}{2\pi}\left[-i\omega\widetilde{x}(\omega)+\widetilde{h}_x(\omega)\right]\nu(\omega)^{-1}\left[i\omega\widetilde{x}(-\omega)+\widetilde{h}_x(-\omega)\right]\right\}\\
&\propto \exp\left\{-\frac{1}{2}\int_{-\infty}^{+\infty}\frac{d\omega}{2\pi}\left[\omega^2\widetilde{x}(\omega)\widetilde{x}(-\omega)+\widetilde{h}_x(\omega)\widetilde{h}_x(-\omega)\right]\nu(\omega)^{-1}\right.\\
&\qquad\left.+\frac{1}{2}\int_{-\infty}^{+\infty}\frac{d\omega}{2\pi}\, i\omega\left[\widetilde{x}(\omega)\widetilde{h}_x(-\omega)-\widetilde{x}(-\omega)\widetilde{h}_x(\omega)\right]\nu(\omega)^{-1}\right\},
\end{aligned} \tag{35}$$

where we have used the short-hand notation

$$\widetilde{x}(\omega) = \gamma x(\omega) + \psi(\omega)\, x(\omega), \qquad \widetilde{h}_x(\omega) = h_x(\omega) - \phi(\omega)\, x(\omega) + m\omega^2 x(\omega). \tag{36}$$

The first integral in the exponential is now even under the replacement $x(\omega) \to x(-\omega)$, while the second is odd. As a consequence the so called *entropy production* reads:

$$\begin{aligned}
\log\frac{\mathcal{P}\{x\}}{\mathcal{P}\{x^{\mathrm{R}}\}} &= \int_{-\infty}^{+\infty}\frac{d\omega}{2\pi}\, i\omega\left[\widetilde{x}(\omega)\widetilde{h}_x(-\omega)-\widetilde{x}(-\omega)\widetilde{h}_x(\omega)\right]\nu(\omega)^{-1}\\
&= -\int_{-\infty}^{+\infty}\frac{d\omega}{\pi}\,\omega\,\mathrm{Im}\left[\widetilde{x}(\omega)\widetilde{h}_x(-\omega)\right]\nu(\omega)^{-1}\\
&= -\int_{-\infty}^{+\infty}\frac{d\omega}{\pi}\,\omega\,\mathrm{Im}\left[x(\omega)h_x(-\omega)\right]\left[\gamma+\psi(\omega)\right]\nu(\omega)^{-1},
\end{aligned} \tag{37}$$

since the term proportional to $x(\omega)\, x(-\omega)$ is projected out when one takes the imaginary part. Note that if h_x is linear in x then the entropy production vanishes for all trajectories. Taking the average over all trajectories, weighted with (25), one gets the average entropy production

$$\left\langle\log\frac{\mathcal{P}\{x\}}{\mathcal{P}\{x^{\mathrm{R}}\}}\right\rangle = -\int_{-\infty}^{+\infty}\frac{d\omega}{\pi}\,\omega\left\langle\mathrm{Im}\left[x(\omega)h_x(-\omega)\right]\right\rangle\left[\gamma+\psi(\omega)\right]\nu(\omega)^{-1}. \tag{38}$$

It is easy to realize, that, in the linear case, $h_x(\omega) \propto x(\omega)$ and (37) vanishes, as discussed in Sect. 3.3.2.

In a general non-equilibrium set up, the entropy production grows linearly in time, namely for large observation time T

$$\left\langle \log \frac{\mathcal{P}\{x\}}{\mathcal{P}\{x^{\mathrm{R}}\}} \right\rangle \sim \sigma T \qquad \text{for } T \gg 1, \tag{39}$$

where σ is the entropy production rate. Formally one can define

$$\left\langle \log \frac{\mathcal{P}\{x\}}{\mathcal{P}\{x^{\mathrm{R}}\}} \right\rangle = \int_{-\infty}^{t} \sigma(s) ds. \tag{40}$$

With the definition

$$K(t) = \int_{-\infty}^{\infty} \frac{d\omega}{2\pi} [\gamma + \psi(\omega)] \, \nu(\omega)^{-1} \tag{41}$$

and by exploiting the properties of the Fourier Transform and ignoring sub leading terms, from (38) one arrives to the following identification:

$$\sigma_t = \int_{-\infty}^{t} dt' K(t-t') \big[\dot{x}(t) h_x(t') + \dot{x}(t') h_x(t) \big], \tag{42}$$

which coincides with the one derived in [3].

3 How to Generate Time Translational Invariant Colored Noise

Let us suppose that our purpose is to reproduce equation with memory (as, for instance in Eq. (3.78)). By exploiting the idea described in Sect. 3.3, one can starts from the Markovian problem:

$$\begin{aligned} \dot{x} &= -\alpha x + \lambda y + \sqrt{2D_x} \, \xi_x \\ \dot{y} &= -\gamma y + \mu x + \sqrt{2D_y} \, \xi_y. \end{aligned} \tag{43}$$

However, once integrated the second equation of (43) , starting from time t_0 with initial condition y_0 and substituted it in the first one, the expression for the *effective* noise of the variable x is

$$\eta(t) = \lambda y_0 g(t-t_0) + \lambda \sqrt{2D_y} \int_{t_0}^{t} ds g(t-s) \phi_y(s) + \sqrt{2D_x} \phi_x(t). \tag{44}$$

At a first sight it seems that Eq. (44) does not satisfy a time translational condition, namely $\langle \eta(t) \eta(t') \rangle \neq f(t-t')$. In order to show this, let us write down the probability distribution of the noise:

$$P[\eta|y_0] = \int \mathcal{D}\sigma\delta\left[\eta - \lambda y_0 g(t-t_0) - \lambda\sqrt{2D_y}\int_{t_0}^{t} ds g(t-s)\phi_y(s) - \sqrt{2D_x}\phi_x(t)\right]. \tag{45}$$

By introducing the *hat* variable $\hat{\eta}(t)$ and by exploiting the integral representation of the delta function, one has

$$P[\eta|y_0] = \int \mathcal{D}\hat{\eta}\mathcal{D}\phi_x\mathcal{D}\phi_y A_\eta B_{\phi_x} C_{\phi_y}, \tag{46}$$

where we have introduced the following notations:

$$A_\eta = \exp\left\{\int_{t_0}^{t_1} dt\, i\hat{\eta}(t)[\eta(t) - \lambda y_0 g(t-t_0)]\right\} \tag{47}$$

$$B_{\phi_x} = \exp\left\{-i\sqrt{2D_x}\int_{t_0}^{t_1} dt\hat{\eta}(t)\phi_x\right\} \tag{48}$$

$$C_{\phi_y} = \exp\left\{-i\lambda\sqrt{2D_y}\int_{t_0}^{t_1} dt\int_{t_0}^{t_1} dt'\hat{\eta}(t')g(t'-t)\phi_y(t)\right\}. \tag{49}$$

Now we use the identity $\langle e^{\lambda x}\rangle = e^{\frac{1}{2}\lambda^2\langle x^2\rangle}$, which is valid for Gaussian integrals, obtaining

$$\int \mathcal{D}\phi_x P[\phi_x]B_{\phi_x} = \exp\left\{-D_x\int_{t_0}^{t_1}\hat{\eta}^2(t)\right\} \tag{50}$$

$$\int \mathcal{D}\phi_y P[\phi_y]C_{\phi_y} = \exp\left\{-\lambda^2 D_y\int_{t_0}^{t_1} dt\int_{t_0}^{t_1} dt'\hat{\eta}(t)\Delta(t,t')\hat{\eta}(t')\right\}, \tag{51}$$

where we have introduced

$$\Delta(t,t') \equiv \int_{t_0}^{t_1} dt' g(t-t'')g(t'-t'') \tag{52}$$

$$= \frac{1}{2\gamma}\left[e^{-\gamma|t-t'|} - e^{-\gamma|t+t'-2t_0|}\right]. \tag{53}$$

Finally, by integrating over the $\hat{\eta}$ the following *Onsager-Machlup* probability distribution is obtained

$$P[\eta|y_0] = \exp\left\{-\frac{1}{2}\int_{t_0}^{t_1} dt\int_{t_0}^{t_1} dt' F[\eta(t)]\nu(t,t')F[\eta(t')]\right\} \tag{54}$$

with

$$F[\eta(t)] \equiv \eta(t) - \lambda y_0 g(t-t_0) \tag{55}$$

$$\nu^{-1}(t,t') = 2D_x\delta(t-t') + \frac{\lambda^2 D_y}{\gamma}\left[e^{-\gamma|t-t'|} - e^{-\gamma(t+t'-2t_0)}\right]. \tag{56}$$

As expected, expression (56) is not of the requested form: the autocorrelation is not time translational invariant and dependence of the initial condition is explicit. However, one can choose the initial condition y_0 randomly with distribution P_0. Then, the final expression for the distribution of the noise is obtained by integrating over the initial condition:

$$P[\eta] = \int dy_0 P_0(y_0) P[\eta|y_0]. \tag{57}$$

By choosing P_0 of the Gaussian form with zero mean and variance $\sigma^2 = \frac{D_y}{\gamma}$, after some calculations, one obtains a colored gaussian process whose correlation is time translational invariant, namely

$$P[\eta] = \exp\left\{-\frac{1}{2}\int_{t_0}^{t_1} dt \int_{t_0}^{t_1} dt' \eta(t)\nu(t,t')\eta(t')\right\} \tag{58}$$

with

$$\nu^{-1}(t,t') = 2D_x\delta(t-t') + \frac{\lambda^2 D_y}{\gamma} e^{-\gamma|t-t'|}. \tag{59}$$

In conclusion, from this example we learn the correct procedure to reproduce a colored noise with correlation (59) by using an auxiliary variable. In order to obtain it, it is necessary to choose the initial condition y_0 from a specific random distribution.

4 Calculation of First Two Coefficients of the Kramers-Moyal Expansion

For larger generality (whose motivation is discussed in the Conclusions), in this Appendix we discuss the case where the gas surrounding the intruder may have a non-zero average $\mathbf{u}$[2]:

$$p(\mathbf{v}) = \frac{1}{\sqrt{(2\pi T_g/m)^d}} \exp\left[-\frac{m(\mathbf{v}-\mathbf{u})^2}{2T_g}\right], \tag{60}$$

which is a simple task involving only the definition of new shifted variables

$$\mathbf{c} = \mathbf{V} - \mathbf{u} \tag{61}$$

[2] Note that in all the cases discussed in the main text, we have always taken $\mathbf{u} = 0$.

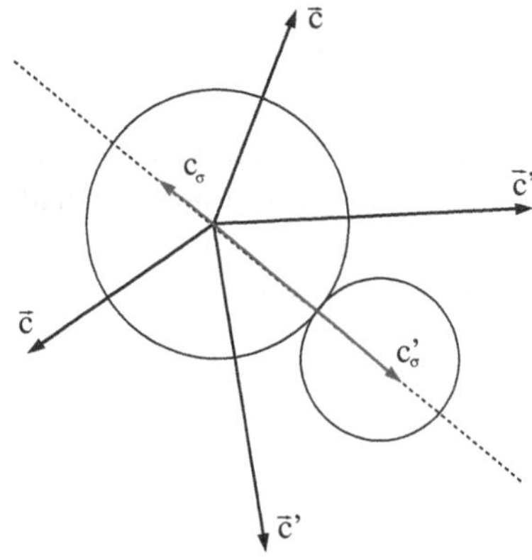

Fig. 1 An example for the change of variables $(c'_x, c'_y) \to (c_\sigma, c'_\sigma)$, introduced in Eq. (64). Such change of variable, when inverted, has two possible determinations: in this example both represented vectors $\mathbf{c}'$ yield the same (c_σ, c'_σ)

$$\mathbf{c}' = \mathbf{V}' - \mathbf{u}. \tag{62}$$

We are interested in computing

$$\begin{aligned} D_i^{(1)}(\mathbf{V}) &= \int d\mathbf{V}'(V_i' - V_i)W_{tr}(\mathbf{V}'|\mathbf{V}) \\ &= \int d\mathbf{c}'(c_i' - c_i)\chi\frac{1}{\sqrt{2\pi T_g/m}k(\epsilon)^2} \\ &\times \exp\left\{-m\left[c'_\sigma + (k(\epsilon)-1)c_\sigma\right]^2/(2T_g k(\epsilon)^2)\right\}. \end{aligned} \tag{63}$$

In order to perform the integral, we make the following change of variables (see Fig. 1 for an example)

$$\begin{aligned} c_\sigma &= c_x\frac{c'_x - c_x}{\sqrt{(c'_x - c_x)^2 + (c'_y - c_y)^2}} + c_y\frac{c'_y - c_y}{\sqrt{(c'_x - c_x)^2 + (c'_y - c_y)^2}} \\ c'_\sigma &= c'_x\frac{c'_x - c_x}{\sqrt{(c'_x - c_x)^2 + (c'_y - c_y)^2}} + c'_y\frac{c'_y - c_y}{\sqrt{(c'_x - c_x)^2 + (c'_y - c_y)^2}} \end{aligned} \tag{64}$$

which implies

$$d\mathbf{c}' = dc'_x dc'_y \to dc_\sigma dc'_\sigma |J|, \tag{65}$$

where

$$|J| = \frac{|c'_\sigma - c_\sigma|}{\sqrt{c_x^2 + c_y^2 - c_\sigma^2}}\Theta(c_x^2 + c_y^2 - c_\sigma^2) \tag{66}$$

is the Jacobian of the transformation. The collision rate is then

$$r(\mathbf{V}) = \chi\sqrt{\frac{\pi}{2T_g/m}} e^{-\frac{mc^2}{4T_g}} \left[(c^2 + 2T_g/m) I_0 \left(\frac{mc^2}{4T_g} \right) + c^2 I_1 \left(\frac{mc^2}{4T_g} \right) \right], \tag{67}$$

where $I_n(x)$ are the modified Bessel functions. For $D_i^{(1)}$ we can write

$$\begin{aligned} D_i^{(1)}(\mathbf{V}) &= \chi \int_{-\infty}^{+\infty} dc_\sigma \int_{c_\sigma}^{\infty} dc'_\sigma (c'_i - c_i)|J| \frac{1}{\sqrt{2\pi T_g/mk(\epsilon)^2}} \\ &\quad \times \exp\left\{ -m \left[c'_\sigma + (k(\epsilon) - 1)c_\sigma \right]^2 / (2T_g k(\epsilon)^2) \right\} \\ &= \chi \int_{-c}^{+c} dc_\sigma \int_{c_\sigma}^{\infty} dc'_\sigma (c'_i - c_i) \frac{c'_\sigma - c_\sigma}{\sqrt{c^2 - c_\sigma^2}} \\ &\quad \times \frac{1}{\sqrt{2\pi T_g/mk(\epsilon)^2}} \exp\left\{ -m \left[c'_\sigma + (k(\epsilon) - 1)c_\sigma \right]^2 / (2T_g k(\epsilon)^2) \right\}, \end{aligned} \tag{68}$$

where we have enforced the constraint of the theta function, namely $c_\sigma \in (-c, +c)$, with $c = \sqrt{c_x^2 + c_y^2}$. Notice that the integral in dc'_σ is lower bounded by the condition $c'_\sigma \geq c_\sigma$ which follows from the definition of c_σ. In order to compute the integral, we have to invert the transformation (64). That yields two determinations for the variables c'_x and c'_y (see Fig. 1)

$$(A) \begin{cases} c'_x - c_x = \frac{c'_\sigma - c_\sigma}{c^2} \left(c_\sigma c_x + c_y Sign(c_x) \sqrt{c^2 - c_\sigma^2} \right) \\ c'_y - c_y = \frac{c'_\sigma - c_\sigma}{c^2} \left(c_\sigma c_y - c_x Sign(c_x) \sqrt{c^2 - c_\sigma^2} \right) \end{cases}$$

$$(B) \begin{cases} c'_x - c_x = \frac{c'_\sigma - c_\sigma}{c^2} \left(c_\sigma c_x - c_y Sign(c_x) \sqrt{c^2 - c_\sigma^2} \right) \\ c'_y - c_y = \frac{c'_\sigma - c_\sigma}{c^2} \left(c_\sigma c_y + c_x Sign(c_x) \sqrt{c^2 - c_\sigma^2} \right) \end{cases}$$

Then the integral (68) can be written as

$$\begin{aligned} D_x^{(1)}(\mathbf{V}) &= \frac{1}{l_0} \int_{-c}^{c} dc_\sigma \int_{c_\sigma}^{\infty} dc'_\sigma \left[(c'_x - c_x)^{(A)} + (c'_x - c_x)^{(B)} \right] |J| \\ &\quad \times \frac{1}{\sqrt{2\pi T_g/mk(\epsilon)^2}} \exp\left\{ -m \left[c'_\sigma + (k(\epsilon) - 1)c_\sigma \right]^2 / (2T_g k(\epsilon)^2) \right\}, \end{aligned} \tag{69}$$

yielding

$$D_x^{(1)} = -\frac{2}{3}\frac{1}{l_0}k(\epsilon)\sqrt{\frac{m\pi}{2T_g}}c_x e^{-\frac{mc^2}{4T_g}}\left[(c^2+3T_g/m)I_0\left(\frac{mc^2}{4T_g}\right)+(c^2+T_g/m)I_1\left(\frac{mc^2}{4T_g}\right)\right],$$
$$D_y^{(1)} = -\frac{2}{3}\frac{1}{l_0}k(\epsilon)\sqrt{\frac{m\pi}{2T_g}}c_y e^{-\frac{mc^2}{4T_g}}\left[(c^2+3T_g/m)I_0\left(\frac{mc^2}{4T_g}\right)+(c^2+T_g/m)I_1\left(\frac{mc^2}{4T_g}\right)\right]. \tag{70}$$

Analogously, for the coefficients $D_{ij}^{(2)}$ one obtains

$$\begin{aligned} D_{xx}^{(2)}(\mathbf{V}) &= \frac{1}{2}\frac{1}{l_0}\int_{-c}^{c}dc_\sigma\int_{c_\sigma}^{\infty}dc'_\sigma\left[\left((c'_x-c_x)^{(A)}\right)^2+\left((c'_x-c_x)^{(B)}\right)^2\right]|J| \\ &\quad\times\frac{1}{\sqrt{2\pi T_g/mk(\epsilon)^2}}\exp\left\{-m\left[c'_\sigma+(k(\epsilon)-1)c_\sigma\right]^2/(2T_gk(\epsilon)^2)\right\} \\ &= \frac{1}{2}\frac{1}{l_0}\frac{k(\epsilon)^2}{15}\sqrt{\frac{2m\pi}{T_g}}e^{-\frac{mc^2}{4T_g}} \\ &\quad\times\left\{\left[c^2(4c_x^2+c_y^2)+3T_g(7c_x^2+3c_y^2)/m+15T_g^2/m^2\right]I_0\left(\frac{mc^2}{4T_g}\right)\right. \\ &\quad\left.+\left[c^2(4c_x^2+c_y^2)+T_g(13c_x^2+7c_y^2)/m+3T_g^2/m^2\frac{-c_x^2+c_y^2}{c^2}\right]I_1\left(\frac{mc^2}{4T_g}\right)\right\}, \end{aligned} \tag{71}$$

$$\begin{aligned} D_{xy}^{(2)}(\mathbf{V}) &= \frac{1}{2}\frac{1}{l_0}\int_{-c}^{c}dc_\sigma\int_{c_\sigma}^{\infty}dc'_\sigma\left[(c'_x-c_x)^{(A)}(c'_y-c_y)^{(A)}+(c'_x-c_x)^{(B)}(c'_y-c_y)^{(B)}\right]|J| \\ &\quad\times\frac{1}{\sqrt{2\pi T_g/mk(\epsilon)^2}}\exp\left\{-m\left[c'_\sigma+(k(\epsilon)-1)c_\sigma\right]^2/(2T_gk(\epsilon)^2)\right\} \\ &= \frac{1}{2}\frac{1}{l_0}\frac{k(\epsilon)^2}{5}\sqrt{\frac{2m\pi}{T_g}}e^{-\frac{mc^2}{4T_g}}c_xc_y \\ &\quad\times\left[(c^2+4T_g/m)I_0\left(\frac{mc^2}{4T_g}\right)+\frac{c^4+2c^2T_g/m-2T_g^2/m^2}{c^2}I_1\left(\frac{mc^2}{4T_g}\right)\right]. \end{aligned} \tag{72}$$

Then we introduce the rescaled variables

$$q_x = \frac{c_x}{\sqrt{T_g/m}}\epsilon^{-1} \qquad q_y = \frac{c_y}{\sqrt{T_g/m}}\epsilon^{-1}, \tag{73}$$

obtaining

$$\begin{aligned}
D_x^{(1)}(\mathbf{V}) =& -\frac{2}{3}\frac{1}{l_0}\sqrt{\frac{\pi}{2}\frac{T_g}{m}}q_x k(\epsilon)\epsilon e^{-\frac{\epsilon^2 q^2}{4}} \\
&\times \left[\left(\epsilon^2 q^2+3\right) I_0\left(\frac{\epsilon^2 q^2}{4}\right)+\left(\epsilon^2 q^2+1\right) I_1\left(\frac{\epsilon^2 q^2}{4}\right)\right], \\
D_{xx}^{(2)}(\mathbf{V}) =& \frac{1}{2}\frac{1}{l_0}\frac{1}{15}\sqrt{2\pi}\left(\frac{T_g}{m}\right)^{3/2} k(\epsilon)^2 e^{-\frac{\epsilon^2 q^2}{4}} \\
&\times \left\{\left[\epsilon^4 q^2(4q_x^2+q_y^2)+3\epsilon^2(7q_x^2+3q_y^2)+15\right] I_0\left(\frac{\epsilon^2 q^2}{4}\right)\right. \\
&+\left.\left[\epsilon^4 q^2(4q_x^2+q_y^2)+\epsilon^2(13q_x^2+7q_y^2)+3\frac{-q_x^2+q_y^2}{q^2}\right] I_1\left(\frac{\epsilon^2 q^2}{4}\right)\right\} \\
D_{xy}^{(2)}(\mathbf{V}) =& \frac{1}{2}\frac{1}{l_0}\frac{1}{5}\sqrt{2\pi}\left(\frac{T_g}{m}\right)^{3/2} q_x q_y k(\epsilon)^2 \epsilon^2 e^{-\frac{\epsilon^2 q^2}{4}} \\
&\times \left[\left(\epsilon^2 q^2+4\right) I_0\left(\frac{\epsilon^2 q^2}{4}\right)+\left(\frac{\epsilon^4 q^4+2\epsilon^2 q^2-2}{\epsilon^2 q^2}\right) I_1\left(\frac{\epsilon^2 q^2}{4}\right)\right].
\end{aligned} \tag{74}$$

Up to this last results we have not introduced any small ϵ approximation. The next step consists in assuming that $q \sim \mathcal{O}(1)$ with respect to ϵ, which is equivalent to assume that $c^2 \sim T_g/M$: this assumption must be compared to its consequences, in particular to Eq. (4.27); the assumption is good for not too small values of α and for $\gamma_g \gg \gamma_b$, i.e. when $T_{tr} \sim T_g$. When this is the case, expanding in ϵ and using that $I_0(x) \sim 1 + x^2/4$ and $I_1(x) \sim x/2$ for small x, one finds Eq. (4.22).

References

1. Baiesi, M., Boksenbojm, E., Maes, C., Wynants, B.: Nonequilibrium linear response for markov dynamics. II: Inertial dynamics. J. Stat. Phys. **139**, 492 (2010)
2. Baiesi, M., Maes, C., Wynants, B.: Nonequilibrium linear response for markov dynamics. I: Jump processes and overdamped diffusions. J. Stat. Phys. **137**, 1094 (2009)
3. Zamponi, F., Bonetto, F., Cugliandolo, L.F., Kurchan, J.: A fluctuation theorem for non-equilibrium relaxational systems driven by external forces. J. Stat. Mech. P09013 (2005)

Biography of the Author

Dr. Dario Villamaina made both his undergraduated and PhD studies in Physics at "La Sapienza" University of Rome with highest marks. From 2011 to 2013 has got a CNRS post-doctoral fellow at the Laboratoire de Physique Théorique et Modèles Statistique, at the Université Paris Sud, Orsay, France.

He is currently at the École Normale Superiéure in Paris, where he has won the "Philippe Meyer" post-doctoral fellowship. His research interests range from non-equilibrium statistical mechanics and kinetic theory to statistics of correlated variables and random matrices. He is co-author of more than 15 publications on peer-reviewed journals on these subjects.

His PhD thesis has been selected for the Springer Theses collection.

D. Villamaina, *Transport Properties in Non-Equilibrium and Anomalous Systems*,
Springer Theses, DOI: 10.1007/978-3-319-01772-3,

Biography of the Author

Zeitfracht Medien GmbH
Ferdinand-Jühlke-Straße 7
99095 Erfurt, Deutschland
produktsicherheit@kolibri360.de